The Arab Uprisings

This book investigates the role of social groups in mobilizing resources for protests in repressive contexts. In particular, it examines the impact of organizations and informal groups on individual engagement in the protests developed in 2010–2011 in Tunisia, Egypt, and Syria. Empirical analysis draws on a wave of events and protests that took place between 2010 and 2021. It explores how, in repressive contexts, spontaneous groups and more established and formal organizations continuously switch from one form to another, transforming themselves faster than they would do in democratic contexts.

Giuseppe Acconcia is a postdoctoral researcher and lecturer in Political Sociology and Geopolitics of the Middle East at the University of Padua, Department of Political Science, Law, and International Studies (SPGI), Italy. He holds a PhD degree in politics at the University of London (Goldsmiths). His research interests focus on social movements, state, and transformation in the Middle East.

Lorenza Perini is a researcher at the Department of Political Science, Law, and International Studies (SPGI), University of Padua, Italy. She holds a PhD degree in contemporary history and in urban planning. She teaches gender policies and globalization. She is a member of the Equal Opportunities Committee and the Gender Studies Research Centre (CEC) at the same University.

The Arab Uprisings
Protests, Gender and War (2011–2021)

Giuseppe Acconcia and Lorenza Perini

Routledge
Taylor & Francis Group

LONDON AND NEW YORK

First published 2023
by Routledge
4 Park Square, Milton Park, Abingdon, Oxon OX14 4RN

and by Routledge
605 Third Avenue, New York, NY 10158

Routledge is an imprint of the Taylor & Francis Group, an informa business

British Library Cataloguing-in-Publication Data
A catalogue record for this book is available from the British Library

Library of Congress Cataloging-in-Publication Data
A catalog record has been requested for this book

ISBN: 978-1-032-27484-3 (hbk)
ISBN: 978-1-032-27572-7 (pbk)
ISBN: 978-1-003-29335-4 (ebk)

DOI: 10.4324/9781003293354

Typeset in Sabon
by Apex CoVantage, LLC

Contents

Acknowledgments

We thank Katia Pilati for her participation in the definition of the theoretical framework and we also thank David Leone Suber and Henda Chennaoui for their support in conducting interviews in Tunisia that are included in Chapters 1 and 3.

Premise – the book in short

The aim of this book is to investigate the role of social groups and women in mobilizing resources for protests in repressive contexts. The research is divided into two parts, whose aims are to consider some fundamental aspects of the role and organization of social movements and groups of people in situations of oppression and endangerment of their fundamental rights. In particular, as the chapters unfold, attention is focused on the plight of LGBTQIA+ (LGBT hereafter) people in terms of the violation of their primary rights and the reduction (if not erasure) of the right to health for women in refugee camps.

The first part of the book examines the impact of organizations and informal groups on individual engagement in the protests developed in 2010–2011 in Tunisia, Egypt, and Syria. The empirical analysis draws on the following data sources: the second wave of the Arab Barometer (2010–2011), two focus group discussion in Egypt conducted between 2011 and 2015 with members of trade unions and of Popular Committees who had participated in the 2011 protests in Egypt, eight semi-structured interviews conducted in 2017 with workers in Tunisia who had engaged in the 2010–2011 protests, and interviews conducted in January and February 2011 with 100 women in Tunisia as part of a study tackling police violence against women during the uprisings.

The second part of the book revolves mainly around a chapter that examines Syrian women displaced in Lebanon through the lens of their access to both health procedures and the possibility of freely deciding on their motherhood. The research has been conducted using a mixed methodology based on the analysis of the existing literature on the topic, as well as on the analysis of existing qualitative and quantitative research reports, with reference to the international human rights standards provided by the instruments concerning asylum seeking, refugee rights, and reproductive rights. Four in-depth interviews have been carried out to support the qualitative analysis, giving back the women their voices and their agency.

The participation of marginalized people in protest movements and the effect of repression in the Middle East and North Africa (MENA) region are also tackled in this part of the book: 44 semi-structured interviews were conducted with LGBT communities in Egypt, Tunisia, and Turkey, and 14 interviews with Kurdish fighters also shed light on female participation within the armed struggle in Northern Syria.

The findings of the first part of the book show that in both Egypt and Tunisia, protests were neither spontaneous nor fully framed into formal organizations, and informal and spontaneous groups were strictly interconnected in sustaining protests. In Egypt, established Islamic charity networks provided the structural basis for Popular Committees to engage in the 2011 protests, while the initially spontaneous workers' groups, institutionalized through the legalization of the Egyptian Federation of Independent Trade Unions (EFITU), were crucial to the nationwide protests that occurred throughout 2011. In Tunisia, the major trade union *Union Générale Tunisienne du Travail* (UGTT) was essential to mobilizing workers in the initial stages of the protests, but it was backed by informal and spontaneous groups of workers during the process of protest diffusion. In the context of war in Northern Syria between 2013 and 2016, with the emergence of a very diverse range of jihadist groups, including Islamic State in Iraq and Syria (ISIS), the participants within those Popular Committees felt the need to be involved in direct action, including the armed struggle, in order to protect their neighborhoods and substitute for the constant absence of security personnel. Hence, the results illustrate that the 2010–2011 Tunisian and Egyptian uprisings were well grounded in intermediate mobilizing structures capable of surviving in the interstices of an authoritarian context. They suggest consideration that in repressive contexts, spontaneous groups and more established and formal organizations continuously switch from one form to another, overlap, and transform themselves faster than they would in democratic contexts.

In the second part of the book, the analysis of the condition of Syrian refugee women in Lebanon aims to shed light on the fact that, with no or limited access to technologies and means of communication, they had to struggle very hard to find not only survival strategies but also a voice, a possibility of affirming their capacity of agency, primarily due to the media misrepresentation of them as victims in perpetual need and waiting for external help, and largely as part of the "women with children" framework, thus overlooking their entangled personal histories. The principal outcome of this study is the fact that the utilization of the social ecological model has given us back the idea that, as with the violence against women in the study led by Heise (1998), violation of reproductive rights and the right to parenthood among the Syrian refugees in Lebanon is not only connectable to individual wrongs or a cultural vacuum, but perpetrated on many different levels, among which women find spaces and measures to put forth their personal will, despite their extremely negative framing as "helpless" and "undeveloped" women that is so pervasive in the western representation of people living in the MENA region.

Although in the previous example, women are portrayed as in a perpetual state of victimhood, in the chapter on the Kurdish struggle for independence, women participants in grassroots protests acted as guerrilla fighters. In the context of a full-scale war in Northern Syria, these grassroots mobilization participants felt the need to be involved in direct action, including the armed struggle, in order to protect their neighborhoods and to substitute the constant absence of security personnel with a salient presence of female fighters.

In this book, where possible, we have tried to highlight and report, also with a different graphic style, the words of the people we have interviewed or whose testimony we have reported. The voices of those who are usually said by others have been our only guide, and we are very keen that those who will read this book can hear and distinguish their sound.

Reference

Heise, L. L. (1998). Violence against women: An integrated, ecological framework. *Violence Against Women, 4*(3), 262–290.

Introduction

Multiple representations of the Arab Spring

There are many different narratives surrounding the events related to the 2011 Arab Uprisings and the subsequent backlash that led to the 2013 military coup in Egypt and civil wars in Syria and Libya. Even 10 years after the first protests, a soldier, a policeman, a supporter of the dissolved National Democratic Party (NDP), one affiliated with the Muslim Brotherhood, mobilized women, LGBT supporters, liberals, and left-wing activists would all give very diverse, and often opposite, accounts of the major reasons for mobilization and demobilization during the demonstrations that took place in urban and peripheral settings between 2011 and 2020 in the MENA region.

The protests that erupted in Tunisia in December 2010 and in Egypt in January 2011 were the product of large cross-class coalitions in which young people and students joined middle-class professionals, government employees, workers, housewives, and the unemployed. Most scholars agree that these protests were spontaneous, often relying on informal ties among friends or neighbors and sustained by online ties. Scholars have also argued that these protests lacked coordination by major social movement organizations (SMOs). However, not all of the evidence is consistent with such claims. For instance, in the case of Tunisia, local groups of the major trade union, UGTT, were crucial to the emerging protests.

Mobilization and demobilization processes

Since the 1970s, the mainstream literature on social movements has argued that SMOs are among the most critical groups capable of mobilizing resources for collective actions (McCarthy & Zald 1977; Diani 2015). This basic tenet has been largely revised due to the diffusion of ICTs (Bennett & Segerberg 2013), as clarified by the claim that it is possible to "organize without organizations." This has appeared even truer for collective actions taking place in repressive contexts; the Arab Uprisings were welcomed as the "Facebook revolutions." Such protests have hence been examined as

DOI: 10.4324/9781003293354-1

spontaneous events supported by ICTs (Lim 2012; Howard & Hussain 2013; Steinert-Threlkeld 2017; Hamanaka 2020), rather than as a result of mobilization by SMOs, as the traditional literature on social movements would have sustained. In fact, the oppositional space in which SMOs can operate is narrow under repressive conditions, and informal networks and loosely structured social groups, often supported by the use of ICTs, are likelier to mobilize resources for political engagement (Clark 2004; Bayat 2010; Duboc 2011; Trejo 2012).

Despite the emphasis, especially by the media, on the spontaneity and sudden eruption of protests in January 2011, the Arab Uprisings have also been examined as part of a long-term revolutionary process that started long before the 2011 upheaval (Achcar 2013; Abdelrahman 2015). For instance, protests spread across Egypt throughout the early 2000s in various campaigns, including the pro-Palestine university mobilizations in 2000, protests against the US invasion of Iraq in 2003, the movements around *Kifaya* in 2004, the march of the judges for independence of the judiciary system in 2006, and workers' protests beginning in 2008. In line with this, and despite the general emphasis on spontaneity, we aim to focus on the variety of groups and organizations that became mobilizing structures during and in the aftermath of the Egyptian Arab Uprisings. While political organizations under authoritarian regimes may not operate as they do in democratic contexts, both apolitical organizations and informal groups can provide resources for political engagement. This often occurs through a mixture of organized and spontaneous mobilizations (Snow & Moss 2014).

As for the processes of demobilization, we argue that waves of contention end in highly varied and contingent ways. As argued by Koopmans, the range of possible endings is principally unlimited and includes regime replacement through revolution, civil war or foreign intervention, repression, elite closure, reform, institutionalization, cooptation, altered conflict and alliance structures, a new balance of electoral power and changes in government incumbency, or any combination of these (Koopmans 2004: 36). The analysis of waves of contention has long informed the literature on contentious politics and is an important area of study in several fields of social sciences, including social movements, ethnic conflicts, civil wars, and civil resistance (Tarrow 1989; Koopmans 2004; Beissinger 2002; Tilly & Tarrow 2015, to cite a few). Under this framework, however, the ways in which processes of demobilization unfold, and the factors associated with different outcomes of demobilization have been studied far less systematically than the emergence of protests and processes of mobilization and factors associated with the rise of contention (see, however, Davenport 2015). Nonetheless, events occurring in the MENA region after the 2010–2011 uprisings demonstrate that an understanding of how processes of demobilization occur is also of utmost importance. This is especially true for postrevolutionary political transitions, as the repression and violence that may occur in such a phase can often be amplified, further unsettling the political context and eventually contributing to an authoritarian backlash.

Drawing on social movement insights into the role of intermediate structures of resource mobilization and on area studies focused on the MENA region, this book aims to investigate the role of various types of social groups – namely, informal and spontaneous ones like peer groups as well as more formal ones like organizations – in protest engagement. Our aim is to examine the so-called Arab Spring and answer the following research questions: Which types of groups were useful in mobilizing individuals to join the protests that erupted in December 2010 in Tunisia and in January 2011 in Egypt? What were the degree of spontaneity and the role of informal ties compared to the role of more structured organizations during the Arab Uprisings? To what extent did the fragmentation of the challengers' coalition in postrevolutionary Egypt contribute to a counterrevolution, while in Tunisia, the challengers' alliances rooted in the prerevolutionary period lasted throughout the phase of demobilization and supported a democratic transition? What repertoires of action were articulated during mobilization and repression by women, LGBT activists, and refugees?

The book intends to shed light on the grassroots participation of social groups in mobilizing resources for protests in repressive contexts. In particular, it examines the impact of organizations and informal groups on individual engagement in the protests that developed between 2010 and 2020 in Tunisia, Egypt, Turkey, and Syria. The empirical analysis draws on the following data sources: the second wave of the Arab Barometer (2010–2011); two focus groups discussion in Egypt conducted between 2011 and 2015 with members of trade unions and of Popular Committees who had participated in the 2011 protests in Egypt; eight semi-structured interviews conducted in 2017 with workers in Tunisia who had engaged in the 2010–2011 protests; interviews conducted in January and February 2011 with 100 women in Tunisia as part of a study tackling police violence against women during the Tunisian uprisings; 44 semi-structured interviews conducted with LGBT communities in Egypt, Tunisia, and Turkey; and 14 interviews with Kurdish fighters in Northern Syria.

The findings analyzed in Chapters 1, 2, and 3 show that in both Egypt and Tunisia, protests were neither spontaneous nor fully organized, as formal organizations and informal and spontaneous groups were strictly interconnected in sustaining protests. In Egypt, established Islamic charity networks provided the structural basis for Popular Committees to engage in the 2011 protests, and the initially spontaneous workers' groups, institutionalized through the legalization of EFITU, were crucial to the nationwide protests that occurred throughout 2011. In Tunisia, the major trade union UGTT was essential to mobilizing workers in the initial stages of protests, but it was backed by informal and spontaneous groups of workers during the process of protest diffusion. In this way, the results highlight that the 2010–2011 Tunisian and Egyptian uprisings were well grounded in intermediate mobilizing structures capable of surviving in the interstices of an authoritarian context. They suggest consideration that in repressive contexts, spontaneous groups and more established and formal organizations continuously switch from one form to another, overlap, and transform themselves faster than they would in democratic contexts.

The different representation of women and LGBT activists

In Chapters 4, 5, and 6, the mobilization strategies articulated by female activists in Syria and Rojava, as well as LGBT communities in Egypt, Tunisia, and Turkey, will be analyzed. In particular, Syrian women in refugee camps – with no or limited access to technologies and means of communication – had to struggle very hard to find not only survival strategies, but also a voice, a possibility of affirming their capacity for agency, largely due to their media misrepresentation as victims, persons in perpetual need and waiting for external help, and as part of the "women with children" framework. In doing so, this overlooked the prospect of disentangling their personal histories, needs, and rights, especially since what they are denied is access to health facilities and the possibility of making free choices regarding their bodies; this is not only as a result of an exceptional state of war but as a systemic condition of social oppression, threat, and difficulties. Conversely, in the chapter on the Kurdish struggle for independence, women participants in grassroots protests acted as guerrilla fighters. Thus, the context of war in Northern Syria between 2013 and 2016 witnessed the emergence of a very diverse range of jihadist groups, including ISIS, as both male and female participants within Popular Committees felt the need to be involved in more direct action, including the armed struggle, in order to protect their neighborhoods and to substitute for the perpetual absence of security personnel.

With regard to LGBT individuals in the MENA region, drawing on social movement theories (SMT) and gender studies, this research aims to explore how these individuals mobilized, as well as the roles played by civil society organizations and digital technologies in the development of such mobilizations. In Egypt, Tunisia, and Turkey, LGBT communities have been disproportionately targeted by state and non-state repressive campaigns. In Egypt, LGBT activists have challenged repression thanks to the use of social networks as alternative venues for socialization, while in Tunisia and Turkey, LGBT activists, drawing on more established meso-level mobilizing structures, have built and implemented new strategies with the intention of increasing their cooperation with other political challengers.

References

Abdelrahman, M. (2015). *Egypt's long revolution protest movements and uprisings.* New York, NY: Routledge.

Achcar, G. (2013). *The people want. A radical exploration of the Arab uprising.* London: Saqi Books.

Bayat, A. (2010). *Life as politics – How ordinary people change the Middle East.* Amsterdam: Amsterdam University Press.

Beissinger, M. (2002). *Nationalist mobilization and the collapse of the Soviet State.* Cambridge: Cambridge University Press.

Bennett, W. L., & Segerberg, A. (2013). *The logic of connective action: Digital media and the personalization of contentious politics.* New York, NY: Cambridge University Press.

Clark, J. (2004). Islamist women in Yemen: Informal nodes of activism. In Q. Wiktorowitz (Ed.), *Islamic activism: A social movement theory approach* (pp. 164–184). Bloomington, IN: Indiana University Press.

Davenport, C. (2015). *How social movements die: Repression and demobilization of the Republic of New Africa.* New York, NY: Cambridge University Press.

Diani, M. (2015). *The cement of civil society. Studying networks in localities.* New York, NY: Cambridge University Press.

Duboc, M. (2011). Egyptian leftist intellectuals' activism from the margins: Overcoming the mobilization/demobilization dichotomy. In J. Beinin & F. Vairel (Eds.), *Social movements, mobilization, and contestation in the Middle East and North Africa* (pp. 61–79). Stanford, CA: Stanford University Press.

Hamanaka, S. (2020). The role of digital media in the 2011 Egyptian revolution. *Democratization, 27*(5), 777–796.

Howard, P. N., & Hussain, M. M. (2013). *Democracy's fourth wave?: Digital media and the Arab spring.* New York, NY: Oxford University Press.

Koopmans, R. (2004). Protest in time and space: The evolution of waves of contention. In D. A. Snow, S. A. Soule, & H. Kriesi (Eds.), *The Blackwell companion to social movements* (pp. 19–46). Oxford: Blackwell.

Lim, M. (2012). Clicks, cabs, and coffee houses: Social media and oppositional movement in Egypt. *Journal of Communication, 62*(2), 231–248.

McCarthy, J. D., & Zald, M. N. (1977). Resource mobilization and social movements: A partial theory. *American Journal of Sociology, 82,* 1212–1241.

Snow, D. A., & Moss, D. M. (2014). Protest on the fly: Toward a theory of spontaneity in the dynamics of protest and social movements. *American Sociological Review, 79*(6), 1122–1143.

Steinert-Threlkeld, Z. C. (2017). Spontaneous collective action: Peripheral mobilization during the Arab spring. *American Political Science Review, 111*(2), 379–403.

Tarrow, S. (1989). *Democracy and disorder: Protest and politics in Italy, 1965–1975.* Oxford and New York, NY: Clarendon Press and Oxford University Press.

Tilly, C., & Tarrow, S. (2015). *Contentious politics* (2nd ed., 2007). New York, NY: Oxford University Press.

Trejo, G. (2012). *Popular movements in autocracies: Religion, repression, and indigenous collective action in Mexico.* New York, NY: Cambridge University Press.

1 Between organization and spontaneity of protests[1]

with Katia Pilati

This chapter investigates the role of social groups in mobilizing resources for protests in repressive contexts. In particular, it examines the impact of organizations and informal groups on individual engagement in the protests developed in 2010 in Tunisia and in 2011 in Egypt.

Findings show that in both Egypt and Tunisia protests were neither spontaneous nor fully organized as formal organizations and informal and spontaneous groups strictly interconnected in sustaining protests. In Egypt, established Islamic charity networks provided the structural basis for Popular Committees to engage in the 2011 protests and the initially spontaneous workers' groups, institutionalized through the legalization of EFITU, were crucial for national wide protests occurred throughout 2011.

In Tunisia, the major trade union UGTT was essential for mobilizing workers in the initial stages of protests but was backed by informal and spontaneous groups of workers during the process of protest diffusion.

Introduction

Numerous hypotheses have been put forward to explain the development of protests in repressive contexts. Many studies have adopted a state-centered or political-process approach that treats repression as a crucial dimension accounting for shifting political opportunities for collective actions (Almeida 2003; Beck 2014; Brockett 1991; Davenport et al. 2005; Goldstone & Tilly 2001; Tilly 1978). These studies have advanced a direct relationship between repression and protest although no consistent evidence on the hypothesis linking repression and protests has been provided (Carey 2009). In an endeavor better to specify the effects of repression on protest mobilization, scholars have suggested that the impact of repression on protests is conditional on intermediate mobilizing structures, that is, on the social groups that can operate in repressive contexts and that are able to mobilize the necessary resources for collective action (Trejo 2012: 84; Kadivar 2013). On the one hand, some scholars have argued that organizations – formal groups implying decisions about membership criteria, rule, hierarchy, monitoring of behaviors and allocation of incentives and sanctions (Ahrne & Brunsson

DOI: 10.4324/9781003293354-2

2011: 86) – play a relevant role not only in democratic contexts for which most evidence has been provided, so far, but also under repressive conditions (Trejo 2012). On the other hand, scholars have argued that informal groups – with no stable structure, where members' roles, positions, and behaviors are not defined by fixed and rules as in organizations, and whose actions often concern daily practices and individual experience (Melucci 1996) – are more crucial in such contexts (Bayat 2010; Pfaff 1996). In the following paragraphs, we discuss in more detail both the roles of organizations and of informal groups in shaping protests in repressive contexts.

The role of organizations

In repressive contexts, the free space available for oppositional social groups is highly restricted (Clark 2004: 170; Moghadam & Gheytanchi 2010; Pilati 2011, 2016; Tchaïcha & Arfaoui 2017; Trejo 2012; Wiktorowicz 2004). Opposition parties are outlawed, and independent social organizations are severely regulated; citizens are often organized into government-controlled associations linked to the ruling party; universities, newspapers, organizations such as workers and farmers' associations, and women and youth rights' groups are often the objects of frequent attacks by ruling governments. SMOs, the most crucial mobilizing structures for collective actions in democratic countries (Davis et al. 2005; Diani 2015; McCarthy & Zald 1977), need to adapt to a constraining environment limiting their usual activities. Under authoritarian rules, SMOs have indeed to adjust and innovate their *repertoire of action* and can radicalize their activities, moderate the repertoire, or transnationalize it (Pilati 2016). Many organizations operating in repressive contexts, rather than being active in explicit political authority-oriented activities, focus on service-delivery and client-oriented actions, (Kriesi 1996) which are not perceived as threatening by political authorities. Organizations focusing on apolitical activities – charities, sport organizations, football fans groups, literacy clubs, or religious organizations – often represent the only free space accessible to individuals in repressive contexts and can become places where broader processes of political socialization take place (Bayat 2010; Clark 2004; Dorsey 2012; Tekeli 1995).

A few mechanisms have been highlighted by social movement scholars to account for the mobilizing role of these organizations. Service-delivery organizations indeed facilitate social and recreational activities, which are places where people discuss and reinforce their sociability networks. In addition, as most voluntary organizations in Europe and in the USA (Verba et al. 1995), they convey resources such as leadership, skills related to the management of collective events, group coordination, and information channels. Next to resource-based mechanisms, these organizations can also provide a rationale for opinions and actions, and can define members' collective identities. In other words, they can provide a "cultural toolkit" of collectively held meanings and symbols used as a collective action frame (McVeigh & Sikkink 2001: 1429). When experiencing repressive measures, apolitical organizations can, therefore, become places for sustaining the creation and intensification of political consciousness and narratives of cultures of resistance. Studies have also shown that football fans groups provide a venue to release anger and frustration, becoming a key institution capable of confronting security force-dominated repressive regimes (Dorsey 2012).

As to the role of organizations in Egypt and Tunisia, there are quite a few evidences on their role before the 2010–2011 protests. The 2008 protests in the textile industry of Mahalla al-Kubra in Egypt had indeed been characterized by wildcat strikes, not supported by the official trade union, the Egyptian Federation of Trade Unions (ETUF), the only recognized trade union in Egypt during Mubarak's regime. However, the local groups of the major trade union in Tunisia, UGTT, have been central during the local protested occurred in 2008 in the Gafsa mining basin (Beinin & Vairel 2011).

Social Islamic organizations and assistance networks, *da'wa*, have, in turn, enabled individuals to become active participants in political life in Egypt (Wickham 2002).

By promoting new values, identities, and commitments, the Islamists had created new motivations for political action. In Egypt, the graduates' embrace of an ideology throughout the nineties was indeed based on framing activism as a "moral obligation" demanding self-sacrifice and unflinching commitment to the cause of religious transformation (Wickham 2002: 148–151). Furthermore, in Egypt, organizations such as *Kifaya* (Enough!), a network striving for reforms and change – including organizations such as Journalists for Change, Doctors for Change, Youth for Change, Workers for Change, and Artists for Change – and the April 6th Youth Movement (A6YM) were also active prior to and during the Arab Spring (Beinin & Vairel 2011). Public gatherings organized from December 2004 to September 2005 in Egypt by *Kifaya* were all possible thanks to the strategy to self-limit the scope of mobilization both in the number of people participating in the organized demonstrations, never exceeding a thousand people, and in the choice of its location, mobilizing in downtown Cairo rather than in densely populated areas. This had enabled *Kifaya* to repeatedly denounce domestic issues, for instance, during public gatherings against President Husni Mubarak's attempts to enact hereditary succession (Beinin & Vairel 2011: 185; Duboc 2011: 32, 61).

In 2005 some members of the Muslim Brotherhood in Alexandria and supporters of the Revolutionary Socialists also formed the National Alliance for Change and Union within the universities. The same activists were later on among those who took part in anti-police riots that broke out after the murder of the young activist Khaled Said in Alexandria in 2010 by a police officer.

In Tunisia, the UGTT was, and currently is, the strongest trade union force in the country, bringing together the middle class with a high concentration in the public sector. One of the strengths of the UGTT has been its political role. Indeed, since independence in 1956, two currents coexist within it: one, controlled by the regime, embodied by what is commonly called the "union bureaucracy"; the other one more inclined to resistance practices – playing a decisive role in the organization of strikes, rallies, and demonstrations – controlling certain federations such as those of education or posts and telecommunications, as well as some regional or local unions. With such characteristics, many social movement groups have found support from the federations and sections of the UGTT.

Next to UGTT, despite the political repression under both Ben Ali and Bourguiba regimes, the Tunisian Association of Democratic Women (ATFD) has also played an important role in the opposition to the regime throughout the nineties and the new millennium. Legalized in 1989, the ATFD focused its struggle against Ben Ali's state feminism, against Islamism, and rising conservatism (Debuysere 2018: 9). While ATFD was concerned with both redistribution and issues related to recognition, the former implied significant support to several protests, which were sometimes undertaken in cooperation with UGTT, like those in Gafsa in 2008, despite the fact that women had been largely underrepresented in the UGTT leadership roles (Debuysere 2018: 10). This provided great legitimacy to ATFD after January 14, 2011, and enabled the organization to play an important role in the subsequent political transition, despite its conflicts with the Islamists.

Following these studies and evidence in Tunisia and Egypt, we can advance that individuals involved in organizations were more likely to join the 2010–2011 protests and demonstrations than those not engaged, despite some differences in relation to the role of trade unions (*Hypothesis 1*).

The role of the informal groups

Other studies show that the mobilization of resources for protests in repressive contexts frequently derives from informal networks and loosely structured social groups, producing dynamics closed to what Bayat has referred to as non-movements (Bayat 2010). Indeed, under authoritarian conditions, places where informal groups meet, where political information becomes diffused, and where collective and shared feelings among activists can develop, are the streets, or the square.

As they have been studied in nonauthoritarian contexts, informal groups are characterized by a latent, loose, and unstable structure and a segmented and reticular network (Melucci 1996).

Due to their flexible, adaptable, and contingent structural nature, informal networks can, therefore, become crucial in repressive contexts (Duboc 2011). Indeed, political claims can be more easily channeled through informal groups of friends, acquaintances, neighbors, getting together in private houses, and cafés, in the street. The hidden and latent structure of these groups not only enables to survive repression but also allows people to directly experience oppositional identities and cultural models and form circuits of solidarity (Melucci 1996: 115).

Thanks to dense and close-knit interactions, informal groups can mobilize primary solidarities, and convincing personal involvement and commitment in the context of rapidly shifting political opportunities. Primary solidarities may nurture the construction of alternative identities, based on the politicization of shared grievances pertaining to private life (Gould 1991; Pfaff 1996: 98). In small and midsize informal groups, where individuals have high levels of trust, loyalties to each other, and strong shared feelings of belonging, expectations of solidarity and participation are possible even under conditions of extreme risk (Gould 1991). Friends and acquaintances may also enable the exchange of political information, political discussion, and political resources and the amount of political discussion occurring in an individual's social network correlates with his or her level of political participation (Klofstad 2011).

Empirical evidence in the MENA region consistently portrays informal groups and ties as significant mobilizing structures in repressive contexts. As argued by Beinin and Vairel (2011: 6–7), most social movements in the MENA region have historically relied on informal networks enabling practices and forms of resistance to bypass the authority. In Egypt, in the 1980s, activism moved to spaces such as those around literary production or journalistic circles (Duboc 2011: 65 ff). Furthermore, workers' protests in Egypt between 2006 and 2009 did not rely on "movement entrepreneurs" or preexisting organizations. With the exception of the support by several labor-oriented NGOs, workers' protests in Egypt mainly relied on irregular face-to-face meetings and mobile telephones, supported by family and neighborhood connections (Beinin 2011: 183).

According to Beinin and Vairel (2011), the 2011 protests in Egypt were indeed possible thanks to personal networks among workers, neighborhood ties, or ties built through workers' children attending the same schools. Clark (2004: 169–170), when examining *nadwas*, that is Qur'anic study or discussion groups where groups of women gather together in homes to read and discuss passages from the Qur'an, highlights three mechanisms of mobilization occurring in such groups: first, they bring Islamist women in regular contact with non-activists; second, they provide a micro-mobilization context in which cognitive liberation and collective attribution can occur; third, they provide the necessary structures of solidarity and interpersonal rewards.

As to Tunisia, woman workers were often organized through informal groups rather than formal membership to UGTT, partly due to the patriarchal attitudes and hierarchy of UGTT as well as to the fact that UGTT did not represent workers who worked informally, a status shared by many women workers (Debuysere 2018).

Following the aforementioned arguments, and against our expectations in hypothesis 1 on the role of organizations, we expect that in contexts of repression, individuals who are part of informal groups – of friends, acquaintances, neighbors, and workmates – are more likely to mobilize resources for protest engagement (*hypothesis 2*). In other words, in repressive contexts, recruitment into protest action is expected to be highly dependent on resources deriving from informal groups.

Methods: data source of quantitative analysis

The empirical study uses data collected in 2010–2011 at the individual level in Egypt and Tunisia, through the second wave of the Arab Barometer survey (Arab Barometer). The dataset contains detailed information on the anti-regime protests occurred in December 2010 in Tunisia, and in January 2011 in Egypt.

The Arab Barometer uses multistage area probability sampling to select respondents, with quota sampling employed in the final stage in several countries[2]. The sample size in Egypt includes 1,144 individuals, and 1,187 individuals are part of the Tunisian sample. These samples represent a national probability sample of adults 18 years and older. Interviews were face-to-face in Arabic and were undertaken between June 16 and July 3, 2011, in Egypt, and between September 30 and October 11, 2011 in Tunisia.

Methods: dependent, independent, and control variables

Dependent variable

The dependent variable of the Egyptian sample considers whether individuals joined the protests against former president Hosni Mubarak between January 25 and February 11, 2011. The dependent variable for the Tunisian sample consists of whether individuals engaged in the protests against former president Zain Al-Abdeen Ben Ali between December 17, 2010, and January 14, 2011. Both of these variables are binary and 1 is assigned to individuals who participated in such protests.

Independent variables

Mobilizing structures. *Formal organizational membership*: To test hypothesis 1 advancing a link between formal membership in organizations and protests, we consider membership in the following types of organizations: a charitable society, a professional association/trade union, a youth/cultural/sport/organization, a family/tribal association (only for the Egyptian sample), and a local development association. *Informal ties*: To test hypothesis 2 on the mobilizing impact of informal groups, we consider whether any friends or acquaintances participated in the protests against former president Mubarak between January 25 and February 11 (for the Egyptian sample) and against former president Ben Ali between December and January 14, 2011 (for the Tunisian sample).[3]

Control variables

Our two hypotheses rely on theories emphasizing the role of intermediate social groups for protests advancing that individuals who are members of organizations and who have friends or acquaintances among protesters are more likely to join the protest. Next to the role of social groups, social movement scholars have also argued that framing processes and an opening political opportunity structure (POS) are crucial for the emergence of protests (McAdam et al. 1996). Following this, we include a proxy for the framing dimension, testing the impact of individual perceptions of the current and future country's economic situation. These variables are aimed to capture the injustice component of collective action (Gamson 2011). We also include attitudes toward authorities, a variable that aims to measure a POS-related dimension. Indeed, according to the political process model, protests are often challenging élites and authorities' *status quo* (Tilly 1978; for more detailed measurement of the POS in the Egyptian case: Barrie & Ketchley 2018). Other control variables test the impact of several sociodemographic and socioeconomic individual characteristics that, according to classical models of political participation in Western countries, are drivers of protests, namely, respondents' sex, civil status, age, educational level attained, and occupational status. Among political attitudes, we include political interest and political information. Finally, we add resources related to the use of the Internet, as well as religious behaviors, namely attending religious lessons in mosques or churches, attending Friday prayers, and reading the Quran.

Data source of qualitative analysis

The qualitative research in Egypt concerns two focus groups, conducted in Sayeda Zeinab (Cairo) and Mahalla al-Kubra (Nile Delta) between 2011 and 2015, involving 16 interviewees.[4] The first focus group tackled seven members of Popular Committees, grassroots groups actively engaged in the 2011 protests, which emerged during the initial stages of the 2011 protests in Egypt active as self-defense groups, check-points, and service-delivery (Hassan 2015). The members involved were male and female unmarried young Egyptians (20–34 years old) with different political and economic backgrounds but mainly coming from middle- and upper-middle class families. They were all living close to Berqet Fil, a small alley parallel to Port Said Street in the popular neighborhood of Sayeda Zeinab. Nine unionized male and female workers were part of the second focus group. They were all living in Mahalla al-Kubra and its outskirts. Two trade unionists and activists, Hamdi Hussein and Gamal Hassanin, acted as gatekeepers in order to select the workers involved in this focus group. In a preliminary stage, Hamdi and Gamal were part of the process for the composition and organization of this specific focus group.

The qualitative research in Tunisia relied, first, on eight semi-structured interviews conducted between February and May 2017 in the neighboring districts of Ben Arous, Medinat Jedida, and Yasminette, in the urban suburbs of Greater Tunis.[5] This area, close to the mercantile port of Rades, is home to one of the most important manufacturing industries of the country, especially for textiles such as denim, mostly used to produce blue jeans. As reported in the 2014 INS census (Institute Nationale de Statistique [INS] 2016), over 25% of the working population in the districts of Medinat Jedida, Yasminette, and Ben Arous is employed in the textile-manufacturing sector, characterizing these districts as areas with some of the highest proportion of the working population employed in the textile industry when compared to the average of the country (18,29%). INS statistics calculate temporary or precariat employees as employed workers.

Most workers employed in the textile sector indeed work on temporary contracts, and often get dismissed and re-employed more times in the same year. Interviews were mostly conducted inside coffee shops or inside private homes, following a "snowball sampling" technique whereby participants would propose other participants.

All interviewees were Tunisians between 25 and 36 years of age, with primary education level, sharing a similar working-class background. All of them had worked at different times in the denim factories of the districts of Ben Arous, Medinat Jedida, and Yasminette. Second, evidence on Tunisia draws on extracts from 100 interviews to women. Interviews had been conducted within the frame of a study, which tackled police violence against women during the protests started in December 2010. Interviews were undertaken between January and February 2011 in the Tunisian town of Kasserine (Zouhour and Nour) and in the village of Thala.

Main findings

Table 1.1 shows the multivariate analysis separately for Egypt (model 1) and for Tunisia (model 2). Given space constraints, in the following discussion, we limit our interpretation to findings addressing our hypotheses. We integrate this analysis with the qualitative interpretation of data drawn from the focus groups in Egypt and the interviews in Tunisia.

With regard to the first hypothesis advancing a link between organizational membership and protests, results show that in Egypt, out of five different types of organizations, only membership in charity organizations increases individuals' probability to join protest in 2011 (cf. model 1 of Table 1.1). While charities do not often engage in political activities and are therefore free spaces capable to operate in repressive contexts, members of these organizations tend to act as political actors. The results of the

Table 1.1 Correlates of engagement in protests; logit regression models (coefficients and standard errors in parenthesis).

	(Model 1)		(Model 2)	
	Egypt		Tunisia	
	b	SE	b	SE
Membership in formal organizations				
Member of charity society	1.370**	(0.445)	0.765	(0.690)
Member of trade union/professional association	−0.830	(0.449)	0.860	(0.523)
Member of youth/cultural/sport association	0.149	(0.567)	0.407	(0.532)
Member of local development association	1.000	(0.974)	0.340	(0.926)
Member of family/tribal association	−1.928	(1.426)		
Informal ties				
Any friends or acquaintances participate in the protests against former president	2.966***	(0.373)	2.228***	(0.291)
Control variables				
Respondent's sex	1.210*	(0.477)	1.307***	(0.288)
Age (range 18–85)	0.003	(0.015)	−0.038**	(0.014)
Being married	−0.241	(0.420)	−0.179	(0.317)
Highest level of education attained	0.085	(0.109)	−0.213*	(0.099)
Occupation (ref: being student)				
Works (no better specified)	–	–	0.145	(0.583)
Retired	0.124	(1.114)	−0.308	(0.765)
Housewife	0.984	(0.932)	−0.666	(0.587)
Unemployed	0.466	(0.917)	−0.147	(0.364)
Employer/director	0.974	(1.097)	1.076	(0.632)
Professional (lawyer, accountant, teacher, doctor, etc.)	2.280*	(0.899)	0.161	(0.662)
Manual laborer	0.868	(1.024)	−0.065	(0.474)
Agricultural worker/owner of a farm	1.506	(0.955)	−0.693	(1.193)
Member of the armed forces/public security	2.601	(2.345)		
Owner of a shop/grocery store	1.483	(0.942)	−0.123	(0.575)
Government employee	1.221	(0.877)	1.102*	(0.518)
Private sector employee	1.090	(0.875)	−0.723	(0.476)
Craftsperson	1.638	(0.935)		
Very interested or interested in politics	−0.152	(0.493)	−0.188	(0.281)
Follows political news very often or often	0.678	(0.580)	0.770*	(0.303)
Reading the Quran always or most of the time	−0.010	(0.380)	0.406	(0.251)
Attends religious lessons in mosques or churches	−0.201	(0.348)	0.709	(0.383)
Attends Friday prayers	−0.125	(0.429)	−0.403	(0.290)
Uses internet at least once a week	0.520	(0.340)	0.601*	(0.283)
Evaluation of current country's economic situation	0.001	(0.200)	0.414**	(0.153)

	(Model 1)		(Model 2)	
	Egypt		Tunisia	
	b	SE	b	SE
Evaluation of future country's economic situation	−0.005	(0.158)	0.051	(0.127)
Supports government's decision	0.192	(0.155)	0.102	(0.130)
Constant	−7.286***	(1.325)	−4.060***	(0.893)
Ll	−191.52		−279.02	
N	1,025		852	

Source: Arab Barometer wave 2, 2010–2011.

Significance levels: *p < .05 **p < .01 *** p < .001.

quantitative analysis in Table 1.1 are integrated by the qualitative analysis that provides evidence for the impact of charities on protests throughout 2011. Indeed, from January 2011 onwards, Popular Committees, rooted in the Islamic charity sector, have been crucial for the continuation of protests. Many members of Popular Committees were in fact supporters of the Muslim Brotherhood and of *Salafi* groups (El-Meehy 2012). In particular, Popular Committees were often rooted within the preexisting networks of Muslim Brotherhood charities, Private Voluntary Organizations (PVO), schools, and hospitals. As the interviewee Mustafa confirmed:

> In Berqet Fil, many participants within the Popular Committees were involved in associations working with the elders or providing social services to the disabled.[6]

Popular Committees were initially informal groups that acted as "self-defense groups – heterogeneous in their tactics, organization, and efficacy – providing a critical response to the security vacuum" (Hassan 2015: 383–386). Indeed, in three days, after the first demonstration in Tahrir Square in Cairo on January 25, 2011, the police began to retreat or apparently disappear from the Egyptian streets. Following this, in a few hours, Popular Committees were quickly organized. "Neighborhood watch brigades, typically led by young men, sprang up to fill the security void as reports of criminal violence mounted" (El-Meehy 2012). As the interviewee Mustafa[7] added, the Popular Committees were spontaneous networks built up on the perfect knowledge of each neighborhood of the Muslim Brotherhood supporters and their capacity to identify any minimum risk:

> Many of the participants in the Popular Committees in my neighborhood were supporters of the Muslim Brotherhood and were already sharing reciprocal relationships of trust or were active in their Private Voluntary Organizations.[8]

Mustafa[9] added that many Popular Committees of his neighborhood were directly forged from the organizational structures of the local branches of the Muslim Brotherhood:

> The Popular Committees have been put in place thanks to the organizational structure of the Muslim Brotherhood and their specific knowledge of the district.

In this sense, Popular Committees have exhibited important continuities with Islamist activism, composed mostly of "upwardly mobile, educated, middle-class professionals" (El-Meehy 2017). Popular Committees participants played a crucial role in promoting individuals' active engagement in politics, including both institutional and noninstitutional politics like protests. According to El-Meehy, in some districts the Committees continued to gather in the spring and summer of 2011 to discuss the main problems of the neighborhood: cleaning streets, fixing water fountains to improve living conditions in the area, and painting buildings but also gradually turned their attention to politics, evolving toward active citizenship (El-Meehy 2012). The Committee's participants were involved in the electoral campaign for the constitutional amendments in the March 2011 referendum, and especially young students or unemployed members of Popular Committees were the first to take part during the continual waves of protest mobilization and the electoral campaigns.

Some of them appeared to be motivated by more conscious revolutionary and secular intentions:

> We wanted a new Constitution. For this reason, we distributed flyers asking to the people to vote No.[10]

Others were motivated by a nationalist, populist, and genuine sense of belonging: "We agreed with the decision of the Muslim Brotherhood to support the call of the army not to make major changes to the then existing Constitution."[11]

Ard al-Lewa's Committee mobilized around the establishment of a park, school and a hospital on 14 feddans of vacant land owned by the Ministry of Religious Endowments (*Awqaf*) in the neighborhood. Next door, the committee in Imbaba organized effective nonpayment campaigns for public services the state had failed to provide, such as garbage collection, while Nahia's Committee constructed an on/off ramp to connect the neighborhood to the ring road.

Those who did take part in Sayeda Zeinab's Popular Committees were active in their own neighborhood during the major episodes of mass mobilization that accompanied the spread of new waves of unrest: the *Fridays of Anger*, during the attacks on the State Security (*Amn el-Dawla*) on April 9, 2011, and the *Baloon Theater* clashes on June 30, 2011.

Some members of the Sayeda Zeinab's Popular Committees extended their mobilization out of the neighborhood. For instance, Ahmed, Midu, and Khaled took part in their first demonstration on the occasion of the Mohammed Mahmud Street clashes of November 2011:

> *This was the first time we went to Tahrir Square. We witnessed the violence of plain clothes policemen, infiltrated within the protests.*[12]
> *At that stage, some of the people of my area went for the first time to Tahrir Square holding Egyptian flags, motivated by their nationalist sense of belonging.*[13]
> *ETUF and independent trade unions' role for protests in Egypt*

Model 1 of Table 1.1 further shows that, in Egypt, other organizations, also those more closely related to political action, such as trade unions and youth organizations, did not act as structures of resources mobilization for the 2011 protests. This confirms the lack of engagement by the official trade unions, Egyptian Federation of Trade Union (ETUF), in the 2011 Egyptian protests. Indeed, the traditional trade unions controlled by Mubarak's regime did not call for labor protests in 2011 even if workers were clearly ready for mass mobilization. Many workers were initially organized in more spontaneous oppositional groups, and workers' committees acting at the local level had no institutional mechanism to compel the ETUF to join the popular movement against Mubarak (Beinin 2011: 107). Spontaneous workers' groups found an institutional form only after the protests broke out, through the creation, on January 30, 2011, of the EFITU, formed thanks to the Center for Trade Union and Workers Services (CTUWS) coordinator, Kamal Abbas, and the Union of Real Estate Tax Authority Workers (RETAU) president, Kamal Abu Eita, along with smaller unions of teachers, health professionals, and retiree associations. In April 2013, the second federation of independent unions was established, the Egyptian Democratic Labour Council (EDLC), convening with 149 unions represented. The pro-regime ETUF continued to support the state institutions despite the fact that the Tahrir Square demonstrations encouraged the workers' groups to mobilize and communicate (Tripp 2013: 160). Early 2011, a nationwide teachers' strike involved half a million workers demanding the *cleansing* (*tathir*) of public institutions of the remnants of the old regime (Hanieh 2013: 169) and in February 2011 in Egypt, 489 strikes occurred.

The workers involved in our second focus group in Egypt, members of independent trade unions, confirmed their early participation in the 2011 protests in Mahalla al-Kubra. Many of them had already been involved in the previous anti-regime mobilizations and strikes.

The participants who were unionized workers demanded better labor conditions and new investments in the textile industries.

> *We were among the hundreds of young people of the revolution gathering in Shon Square in Mahalla al-Kubra.*[14] *We were asking for a better life and human working conditions. We participated during the first*

> *protests after an already long-lasting struggle to overcome the rooted crisis of the Egyptian cotton industry.*[15]

But this interviewee had some previous experience:

> *We have been used for years to go downtown Cairo during mass demonstrations and strikes or close to Mubarak residency in Qasr al-Qobba to demonstrate against his neo-liberal labor policies.*[16]

The role of UGTT and spontaneous workers' groups for protests in Tunisia

The role of trade unions in the Tunisian case is different from the one found for Egypt. Model 2 shows no significant effect of trade union membership on the anti-regime protests erupted in December 2010 in Tunisia. However, the qualitative analysis is clearer in emphasizing the role of UGTT, and its interweaving with more spontaneous groups of workers. In Tunisia, protests were joined by government employees as well as students and unemployed, the two latter making up the largest percentage of protesters. As shown by the qualitative research conducted in Tunisia, testimonies from workers taking part in protests between December 17 and January 14, 2010, indicate a strong reliance on national union mobilization. The local branch of the major Tunisian trade union, UGTT in Ben Arous organized a demonstration in front of its headquarters in the last days of December, to which most members attended.

> *The demonstration calls had been repeated and encouraged by members of the unions, but we attended because we were aware of what was happening in the country.*[17]

Local groups of the UGTT were crucial for the eruption of protests in Ben Arous. Here, groups of workers and unemployed were especially active in the protests despite the fact that their membership to the UGTT was loosely defined due to the presence of many workers with precarious employment and on a temporary contract.

The line between workers and unemployed was therefore blurred in Ben Arous due to the fact that most workers were employed on monthly contracts and no real distinction emerged between workers and unemployed. The organization of workers as a "group" solidified *because of* the call to protest raised by formal organizations of which not all were actually members of.

> *I (recall) the moment I saw the unions' communicate on the phone of a friend of mine. My uncle was sitting in the same café, some tables aside . . ., when he read the message he swore that if the UGTT had called for a*

picket, then also those without a job like me, should have joined on the day.[18]

It was also through informal ties that individuals mobilized as family members, friends, and even neighbors would be encouraged to join protests.

In Yasminette there was not the notion of workers and unemployed, we were all one, the young people of Yasminette.[19]

These groups were supported both by the structure and encouragement provided by the local union branches, and by the informal networks present in the community.

On the day of the strike we were given banners and hats, I still have that hat, as it reminds me of the first day I shouted openly to a policemen.[20]

The key for mobilization in Tunisia was provided by the support of local workers' organizations being integrated by informal ties existing among friends and family members, groups related to either the workplace or through houses, cafés, and the street. Therefore, some groups formed, and members recognized themselves as part of a "group" only during and after the protests. Indeed, groups were both preexisting structures of mobilization, and a result of the political climate emerged out of the December 2010 protests.

Evidence coming from the eight semi-structured interviews with individuals who had participated in the 2011 protests in Tunisia suggests that the level of participation in the protests was higher when protests and strikes were called by formal organizations. Nonetheless, this does not contradict the high degree of spontaneity in the ways in which protesters both joined and performed protest in December 2010 and January 2011 in Ben Arous. As a participant put it:

We would go to demonstrations organized by the local unions in Ben Arous and Medinat Jedida, but when it came to confronting the police, only the youth from the quartier was in the streets.[21]

Such relation – between spontaneous and organized action – can be further examined by considering the role of football fan clubs. Indeed, many young people living in Ben Arous and Yasminette are members of Club Africain's unregulated and outlawed ultras groups, yet having an organizational structure and internal hierarchy capable of managing donation-based funds.

We've been having problems with the police forever. When we went to the streets we just did what we do when we leave the stadium and the police is there.[22]

The flexibility created by the existence of loose memberships to fan groups was critical for workers unions when the first demonstrations erupted in Ben Arous and Yasminette in December 2010.

> *What we needed to confront the police present at the protests in Yasminette was just to get the guys of the Club African down with us and we would have a demonstration. You would see Club African flags waved by people wearing a UGTT union hat!*[23]

Consequently, affiliation to a workers' union or to a football fan group in Ben Arous is not a matter of rigid membership. The fact that such groups facilitated the participation to protests cannot ignore the fact that more loose and spontaneous groups eventually formed and took up the task to initiate and coordinate protest action.

> *After demonstrations or pickets, we split into the same groups we were when we went to the café. . . . The cafés of Ben Arous were the places where we met after and before all protests, these were the places we would go and find other people to join the marches.*[24]

The presence of both formal groups and of informal networks is thus testament of the overall mixed nature of protesters' groups, as many young people who would not usually be part of unions or football fans groups ended up marching with them and clashing with the police in January 2011 in Ben Arous and Yasminette. This nonetheless, informal networks grew strong and became aware of their own existence only after protests were called by workers' unions.

The role of other groups

Model 2 in Table 1.1 further shows that no other organizational membership was significant for the eruption of protests in Tunisia. This result, however, is also due to the limited number of types of organizations considered in the quantitative analysis. In fact, women organizations were also active in big cities and especially in Tunis where the political elite, including a high share of women, had managed to forge an opposition to Ben Ali and Bourguiba regimes. Indeed, in addition to the ATFD, many women were active in political parties and associations as well as in the UGTT, despite not occupying leadership roles. An interview with Amal Dhafouli, a 34-years-old activist in Sidi Bouzid, argues that

> *I am independent but I collaborate a lot with the UGTT and other civil society organizations. At each sit-in or gathering, the UGTT supports us and helps us to open the doors of dialogue with the authorities. But even before the revolution and until today, I have never had to deal with*

women trade unionists in Sidi Bouzid or Tunis. Despite this, women were during the revolution and until today one of the drivers of social movements in the regions.

Female journalists and bloggers played a key role in the 2011 protests. An important example is Lina Ben Meheni, a renown blogger, Nobel nominee, who ensured the coverage of the protest events for several weeks in many languages (Chennaoui & Baraket 2011).

Overall, considering the results both for Tunisia and for Egypt, the impact of engagement in formal organizations is limited. Therefore, hypothesis 1 is only partly met. In contrast, for both Egypt and Tunisia, results of the quantitative analysis in Table 1.1 and the qualitative studies confirm the crucial impact of informal ties, as already discussed, thus supporting hypothesis 2.

More importantly, however, the integration of the quantitative and the qualitative analysis remarks the need to consider the interconnections between spontaneous groups and more stable organizations as, in a repressive context, they overlap, switch from one form to another, and transform themselves faster than they would do in democratic contexts.

Conclusions: between spontaneity and organization

This study has investigated the role of intermediate structures of mobilization in repressive contexts, trying to understand whether the significant effect of groups for the mobilization of resources for protests in democratic contexts differs under authoritarian regimes. To do this, the chapter has integrated the insights of a quantitative analysis with those of a qualitative one. Results suggest that organizations in repressive contexts become mobilizing structures by largely overlapping with informal ties. In particular, protests erupted in December 2010 in Tunisia and in January 2011 in Egypt were coordinated by organizations that often overlapped or emerged from informal ties and closed relationships built every day by neighbors, friends, relatives, and peers. Protests in repressive contexts, therefore, develop in spaces characterized by both spontaneity and organization. An understanding of formal engagement needs to be integrated with an understanding of informal ties which pervaded the Tunisian and Egyptian societies at the time of the uprisings.

Our study, therefore, supports a nuanced vision of the role of organizations for protests in repressive contexts. Indeed, organizations have a flexible form whereby, as exemplified by the Tunisian case, membership is not so strict, and organizations can, at times, incorporate claims by unorganized members. Organizations have also a dynamic form as they are capable to react, and easily adapt to a changing context. The characteristics of groups change fast during the time of political turmoil in repressive contexts. Workers' spontaneous and informal groups in Egypt turned into more stable organizations through processes of institutionalization, as reflected by the legalization of EFITU. Popular Committees developed mostly from members' informal ties built through prior affiliations to the Muslim Brotherhood and to *Salafi* groups. In Tunisia, organizations such as UGTT ignited protests in the initial phase, but informal ties of workers were crucial to widen engagement by workers during the diffusion process.

Protests in 2011 were therefore well grounded on mobilizing structures capable to survive in the interstices of an authoritarian context, namely, small groups of friends and acquaintances integrating, overlapping with, changing into, or deriving from formal, stable, and more organized groups capable to survive the repressive context. This interconnection suggests that neither spontaneity nor coordination and organization prevailed during the Arab uprisings, and that a nuanced vision joining the two different group forms – organizations and informal groups – and their changing dynamics need to be considered.

Notes

1 This chapter has been previously published in a different format in Social Movement Studies. Pilati, K., Acconcia, G., Suber, L. D., Chennaoui, H. (2019). Between organization and spontaneity of protests: The 2010–2011 Egyptian and Tunisian uprisings. *Social Movement Studies, 18*(4).
2 The sampling design included stratification (by governorate and urban-rural) and clustering. Interviews were distributed proportionally to population size. For further methodological details, see the Arab Barometer Project, www.arab-barometer.org/.
3 The sampling design included stratification (by governorate and urban-rural) and clustering. Interviews were distributed proportionally to population size. For further methodological details, see the Arab Barometer Project, www.arab-barometer.org/.
4 For security reasons, we prefer to keep the names of the individuals involved anonymous and provide fictitious names. These focus groups were part of a broader research by Giuseppe Acconcia (2018) conducted between 2011 and 2015 on Popular Committees and Independent Trade Unions in Egypt.
5 The eight semi-structured interviews were part of a broader research by David Leone Suber (2017) and the extracts of the 100 interviews were part of a broader project in which Henda Chennaoui collaborated (Chennaoui & Baraket 2011).
6 Case 1, Interviewee 7, Cairo.
7 Case 1, Interviewee 7, Cairo.
8 Case 1, Interviewee 1, Cairo.
9 Case 1, Interviewee 1, Cairo.
10 Case 1, Interviewees 1, 2, 3, 6, and 7, Cairo.
11 Case 1, Interviewees 4 and 5, Cairo.
12 Case 1, Interviewees 2, 3, and 4, Cairo.
13 Case 1, Interviewees 2, 3, and 4, Cairo.
14 Case 2, Interviewees 3, 4, and 9, Mahalla al-Kubra.
15 Case 2, Interviewee 5, Mahalla al-Kubra.
16 Case 2, Interviewees 1, 2, and 6, Mahalla al-Kubra.
17 Case 3, Interviewee 1, Ben Arous.
18 Case 3, Interviewee 3, Ben Arous.
19 Case 3, Interviewee 7, Yasminette.

20 Case 3, Interviewee 4, Medinat Jedida.
21 Case 3, Interviewee 8, Yasminette.
22 Case 3, Interviewee 7, Yasminette.
23 Case 3, Interviewee 7 and 8, Yasminette.
24 Case 3, Interviewee 2, Ben Arous.

References

Acconcia, G. (2018). *The uprisings in Egypt: Popular committees and independent trade unions* (PhD thesis). London: Goldsmiths College, University of London.

Ahrne, G., & Brunsson, N. (2011). Organization outside organizations: The significance of partial organization. *Organization, 18*(1), 83–104.

Almeida, P. D. (2003). Opportunity organizations and threat-induced contention: Protest waves in authoritarian settings. *American Journal of Sociology, 109*(2), 345–400.

Arab Barometer. (2010–2011). Retrieved from www.arabbarometer.org/waves/arab-barometer-wave-ii/

Barrie, C., & Ketchley, N. (2018). Opportunity without organization: Labor mobilization in Egypt after the 25th January revolution. *Mobilization: An International Quarterly, 23*(2), 181–202.

Bayat, A. (2010). *Life as politics – How ordinary people change the Middle East.* Amsterdam: Amsterdam University Press.

Beck, C. J. (2014). Reflections on the revolutionary wave in 2011. *Theory and Society, 43*(2), 197–223.

Beinin, J. (2011). A workers' social movement on the margins of the global neoliberal order, Egypt 2004–2009. In J. Beinin & F. Vairel (Eds.), *Social movements, mobilization, and contestation in the Middle East and North Africa* (pp. 181–201). Stanford, CA: Stanford University Press.

Beinin, J., & Vairel, F. (Eds.). (2011). *Social movements, mobilization, and contestation in the Middle East and North Africa.* Stanford, CA: Stanford University Press.

Brockett, C. D. (1991). The structure of political opportunities and peasant mobilization in Central America. *Comparative Politics, 23*(3), 253–274.

Carey, S. (2009). *Protest, repression and political regimes: An empirical analysis of Latin America and Sub-Saharan Africa.* New York: Routledge.

Chennaoui, H., & Baraket, S. (2011). *Les Abandonnées de la Révolution – Étude des violences faites aux femmes à Thala et Kasserine lors de la répression de l'insurrection de décembre 2010-janvier 2011.* Uganda: Isis-WICCE Exchange Institute Alumni.

Clark, J. (2004). Islamist women in Yemen: Informal nodes of activism. In Q. Wiktorowitz (Ed.), *Islamic activism: A social movement theory approach* (pp. 164–184). Bloomington, IN: Indiana University Press.

Davenport, C., Hank, J., & Mueller, C. (2005). *Repression and mobilization.* Minneapolis, MN: University of Minnesota Press.

Davis, G. F., McAdam, D., Scott, W. R. E., & Zald, M. N. (2005). *Social movements and organization theory.* Cambridge: Cambridge University Press.

Debuysere, L. (2018). Between feminism and unionism: The struggle for socio-economic dignity of working-class women in pre- and post-uprising Tunisia. *Review of African Political Economy, 45*(155), 25–43.

Diani, M. (2015). *The cement of civil society. Studying networks in localities.* New York, NY: Cambridge University Press.

Dorsey, J. M. (2012). Pitched battles: The role of ultra soccer fans in the Arab spring. *Mobilization: An International Journal, 17*(4), 411–418.

Duboc, M. (2011). Egyptian leftist intellectuals' activism from the margins: Overcoming the mobilization/demobilization dichotomy. In J. Beinin & F. Vairel (Eds.), *Social movements, mobilization, and contestation in the Middle East and North Africa* (pp. 61–79). Stanford, CA: Stanford University Press.

El-Meehy, A. (2012). Egypt's popular committees from moments of madness to NGO dilemmas. *Middle East Research and Information Project, 42*(265). Last time accessed Winter 2016.

El-Meehy, A. (2017). Governance from below. Comparing local experiments in Egypt and Syria after the uprisings, report Arab politics beyond the uprisings. *The Century Foundation.* Retrieved February 10, 2017, from https://tcf.org/content/report/governance-from-below/

Gamson, W. (2011). Arab spring, Israeli summer, and the process of cognitive liberation. *Swiss Political Science Review, 17*(4), 463–468.

Goldstone, J., & Tilly, C. (2001). Threat (and opportunity): Popular action and state response in the dynamic of contentious action. In R. Aminzade, J. Goldstone, D. McAdam, E. Perry, W. Sewell, & S. Tarrow (Eds.), *Silence and voice in the study of contentious politics* (pp. 179–194). Cambridge: Cambridge University Press.

Gould, R. (1991). Multiple networks and mobilization in the Paris Commune, 1871. *American Sociological Review, 56*(6), 716–729.

Hanieh, A. (2013). *Lineages of revolt, issues of contemporary capitalism in the Middle East.* Chicago, IL: Haymarket Books.

Hassan, H. (2015). Extraordinary politics of ordinary people: Explaining the micro dynamics of popular committees in revolutionary Cairo. *International Sociology, 30*(4), 383–400.

Institute Nationale de Statistique. (2016). *Ben Arous: A travers le Recensement Général de la Population et de l'Habitat 2014.* Retrieved from www.ins.nat.tn/fr/publication/ben-arous-%C3%A0-travers-le-recensement-g%C3%A9n%C3%A9ral-de-la-population-et-de-l%E2%80%99habitat-2014

Kadivar, M. A. (2013). Alliances and perception profiles in the Iranian reform movement, 1997 to 2005. *American Sociological Review, 78*(6), 1063–1086.

Klofstad, C. C. (2011). *Civic talk: Peers, politics, and the future of democracy.* Philadelphia, PA: Temple University Press.

Kriesi, H.-P. (1996). The organizational structure of new social movements in a political context. In D. McAdam, J. D. McCarthy, & M. N. Zald (Eds.), *Comparative perspectives on social movements: Political opportunities, mobilizing structures, and cultural framings* (pp. 152–184). Cambridge: Cambridge University Press.

McAdam, D., McCarthy, J. D., & Zald, M. N. (Eds.). (1996). *Comparative perspectives on social movements: Political opportunities, mobilizing structures, and cultural framings.* New York, NY: Cambridge University Press.

McCarthy, J. D., & Zald, M. N. (1977). Resource mobilization and social movements: A partial theory. *American Journal of Sociology, 82*, 1212–1241.

McVeigh, R., & Sikkink, D. (2001). God, politics, and protest: Religious beliefs and the legitimization of contentious tactics. *Social Forces, 4*, 1425–1458.

Melucci, A. (1996). *Challenging codes – Collective action in the information age.* New York, NY: Cambridge University Press.

Moghadam, V., & Gheytanchi, E. (2010). Political opportunities and strategic choices: Comparing feminist campaigns in Morocco and Iran. *Mobilization, 15*(3), 267–288.

Pfaff, S. (1996). Collective identity and informal groups in revolutionary mobilization: East Germany in 1989. *Social Forces, 75*(1), 91–118.

Pilati, K. (2011). Political context, organizational engagement, and protest in African countries. *Mobilization: An International Journal, 16*(3), 351–368.

Pilati, K. (2016). Do organizational structures matter for protests in non-democratic African countries? In E. Y. Alimi, A. Sela, & M. Sznajder (Eds.), *Contention, regimes, and transition – Middle East and North Africa protest in comparative perspective* (pp. 46–72). Oxford: Oxford University Press.

Suber, D. L. (2017). *Failing readmission: If sending migrants back won't work. A case study of Italy and Tunisia.* Berlin: Rosa Luxemburg Stiftung North Africa Office Publication Index.

Tchaïcha, D. J., & Arfaoui, K. (2017). *The Tunisian women's rights movement: From nascent activism to influential power-broking.* London: Routledge.

Tekeli, S. (Ed.). (1995). *Women in modern Turkish society: A reader.* London: Zed Books.

Tilly, C. (1978). *From mobilization to revolution.* Reading, MA: Addison-Wesley.

Trejo, G. (2012). *Popular movements in autocracies: Religion, repression, and indigenous collective action in Mexico.* New York, NY: Cambridge University Press.

Tripp, C. (2013). *The power and the people, paths of resistance in the Middle East.* New York, NY: Cambridge University Press.

Verba, S., Schlozman, K., & Brady, H. (1995). *Voice and equality: Civic voluntarism in American politics.* Cambridge, MA: Harvard University Press.

Wickham, C. R. (2002). *Mobilizing Islam: Religion, activism, and political change in Egypt.* New York, NY: Columbia University Press.

Wiktorowicz, Q. (Ed.). (2004). *Islamic activism – A social movement theory approach.* Bloomington, IN: Indiana University Press.

2 Variety of groups and protests in repressive contexts[1]

with Katia Pilati

This chapter examines the role of various social groups in shaping protests in repressive contexts. The empirical study focuses on the Egyptian uprisings that started in January 2011. The authors use data collected through semi-structured interviews undertaken between 2011 and 2015 with 58 individuals who had all participated in such protests and who were members of various types of organizations. The results show that, in contrast to the arguments highlighting the spontaneous, internet-based nature of the protests that occurred in 2011 in the MENA region, individuals' membership in organizations such as the Muslim Brotherhood, and in more informal groups such as Popular Committees or independent trade unions have been crucial for their engagement in protests. The findings also highlight the flexible and dynamic form of organizations active in repressive contexts, which are capable of reacting and adapting easily to a changing context.

Introduction

In this chapter, we aim to unfold the role of different types of groups, which acted as major structures of mobilization during the 2011 protests in Egypt: the Muslim Brotherhood, and their links with the Popular Committees, self-organized workers' groups and their ties with ensuing institutionalized independent trade unions. By focusing on these groups, we examine the role of established organizations such as the Muslim Brotherhood and trade unions as well as that of less structured and informal groups like Popular Committees or self-organized workers' groups, contending that they were strictly interconnected during and after the 2011 protests. In both cases, one provided the structural basis for the other, and established organizations switch to more informal groups and vice versa.

Our empirical analysis uses data derived from extensive fieldwork notes collected in Egypt between 2011 and 2015. In particular, it draws on 58 semi-structured interviews undertaken between January 2011 and June 2015 in Cairo and Mahalla al-Kubra.

Results highlight the variety of forms of groups, which engaged in the prerevolutionary coalitions with the aim of ousting Mubarak as well as in

DOI: 10.4324/9781003293354-3

the aftermath of the 2011 uprisings, which was characterized by a political transition that culminated in the backlash of an authoritarian regime through the 2013 military coup. As our results will show organizations such as the Muslim Brotherhood provided support for political engagement through other organizational forms, namely loosely organized groups such as Popular Committees. In addition, informal groups such as self-organized workers' groups provided the basis for more established mobilizing structures such as the EFITU to take place once protests in 2011 started.

Variety of mobilizing groups in repressive contexts

Several scholars have approached the analysis of the Arab Uprisings drawing on the literature of social movement that associates the rise of protests to several concepts – including political opportunity structures and mobilizing structures such as organizations and framing processes (McAdam et al. 1996). For instance, by examining protests through the concept of political opportunity structures, Goldstone has argued that "the single best key to where regimes in MENA have been overturned or faced massive rebellions is where personalist regimes have arisen" (Goldstone 2016: 108). Other scholars have put emphasis on the role of mobilizing structures, both offline and online. Among the latter, the role of ICTs such as Facebook or Twitter has been the focus of analyses by various authors (Lim 2012; Howard and Hussain 2013; Steinert-Threlkeld 2017; Hamanaka 2020). Among more classical mobilizing structures, trade unions or Popular Committees have been largely investigated as crucial spaces for the process of resource mobilization (Achcar 2013; Abdelrahman 2015; El-Meehy 2012). In line with these authors, we aim to investigate the variety of social groups, which have fostered the rise of protests in Egypt. In fact, one of the main functions of organizations is the capacity to articulate political demands. Organizations enhance the coordination of collective actions and protests undertaken with the aim of promoting social and political changes (Tilly 1978). As has long been argued (McCarthy & Zald 1977), organizations lower the costs associated with collective actions as they support the coordination of activities thanks to the presence of leadership, and they facilitate sustained and durable social interactions that facilitate the development of collective identities.[2] They also smooth the flows of information and its exchange, mobilize and aggregate resources such as money, provide spaces, equipment, and infrastructures for the implementation of activities, and favor the social and political legitimacy that single actors may not have. Studies have demonstrated that organizational involvement allows people to increase their social capital and to improve their communicative skills, their organizational abilities, and their capacity to manage groups' coordination, therefore facilitating the involvement in collective actions and in political activities (Verba et al. 1995). The aforementioned resources have been highlighted for most SMOs operating under democratic assets. However, in repressive contexts, the oppositional space in which political organizations such as SMOs can operate is narrow, and hinders the possibility for organizations to engage in contentious collective actions. Under repressive conditions, SMOs can adapt or redefine their actions in two different ways: On the one hand, organizations may adhere to associational agendas promoted by authorities that directly serve their political mandates, consequently reinforcing clientelistic behaviors, corruption, and nepotism, and promoting ideals that are not critical of the regimes (Jamal 2007). On the other hand, political organizations may replace their usual activities of lobbying, political networking, or linking with media, to avoid targeted and systematic repressive measures by innovating the repertoire of action, thanks to (1) the radicalization of organizational activities, that is, the use of more confrontational activities, including engagement in violent political actions; (2) the trans-nationalization of organizational activities, that is, their diffusion across national boundaries and states (Tarrow 1996: 52), as is the case of internationally supported nongovernmental organizations (NGOs) focusing on human rights and international law discourses; and (3) the moderation of the organizations' repertoires of action (Pilati et al. 2019). Unlike political organizations, those concentrating on apolitical claims and agendas can more freely operate under authoritative conditions, given that they are likely to be perceived as nonthreatening by the ruling elites (Hinnebusch 2015). Organizations working in service delivery and provision, like charity organizations sustaining the population with employment opportunities, housing, or health assistance, do not represent explicit challenges for authorities. Sometimes, religious organizations can also freely operate under authoritarian regimes (Trejo 2012). Mechanisms accounting for the potential of apolitical organizations to be mobilizing structures include, first, their capacity to facilitate social and recreational activities where people discuss and get to know new people and reinforce their sociability networks. Apolitical organizations are places where broader processes of political socialization also take place (cf. Clark 2004a; Bayat 2010; Dorsey 2012). Second, like most voluntary organizations in Western contexts (Verba et al. 1995), apolitical organizations can provide important resources for contentious collective actions such as leadership, skills related to the management of collective events, group coordination, and dissemination of information otherwise unavailable to individuals. Third, in addition to resource-based mechanisms, these organizations can provide a rationale for opinions and actions as well as for defining members' collective identities (Lichterman 2008). Hence, they can provide a "cultural toolkit" of collectively held meanings and symbols used as a collective action frame (McVeigh & Sikkink 2001: 1429). When experiencing repressive measures, apolitical organizations can therefore become places for sustaining the creation and intensification of a political consciousness and narrative of cultures of resistance.

Scholars have argued that in circumstances where there is limited space for organizations to work, informal groups are crucial sites for the mobilization of resources in repressive contexts (Bayat 2010; Pfaff 1996). By "informal groups" we mean groups with no stable structure, where members' roles, positions, and behaviors are not defined by fixed rules as in organizations, and whose actions often concern daily practices and individual experience (cf. Melucci 1996). Thanks to dense and close-knit interactions, informal groups can mobilize primary solidarities, and convincing personal involvement and commitment in the context of rapidly shifting political opportunities. Solidarities may nurture the construction of alternative identities, based on the politicization of shared grievances pertaining to private life (Gould 1991; Pfaff 1996: 98). In small and midsize informal groups, where individuals have high levels of trust, loyalties to each other, and strong shared feelings of belonging, expectations of solidarity, and participation are possible even under conditions of extreme risk (Gould 1991). Friends and acquaintances may also enable the exchange of political information, political discussion, and political resources. The amount of political discussion occurring in an individual's social network correlates with his or her level of political participation (Klofstad 2011).

As mentioned earlier, ties built on the web may further facilitate collective actions under repressive conditions (Bremer 2012; El-Meehy 2012; Lim 2012; Howard & Hussain 2013; Hamanaka 2020).[3] Below we discuss the variety of organizations and informal groups, as well as their interconnections, which were actively present during and in the aftermath of the protests that erupted in Egypt in January 2011.

Variety of mobilizing groups under Mubarak's regime

Many studies have shown the presence of an important preexisting organizational structure operating throughout the repressive regime under the mandate of Mubarak – who served as the President of Egypt from 1981 to 2011 (Achcar 2013; Della Porta 2014; Abdelrahman 2015). These organizations likely prepared the ground for the protests that erupted in 2011. On the one hand, organizations such as the major governmental trade union in Egypt, the Egyptian Federation of Trade Union (ETUF), had been largely coopted by Mubarak under his mandate. This organization practically played no role in the development and coordination of the January 2011 protests. ETUF was also absent from the coordination of previous protests such as those that erupted in 2008 in Mahalla al-Kubra. On the other hand, other organizations had been crucial for the emergence of many protests that were observed throughout the first decade of the millennium as well as for those that erupted in 2011. *Kifaya* (Enough!), a network striving for reforms and change including organizations such as Journalists for Change, Doctors for Change, Youth for Change, Workers for Change, Artists for Change developed from informal networks among dissenters, and the April 6th Youth Movement (A6YM) had been active both prior to and during the Arab Spring (Beinin 2011).

Public gatherings organized from December 2004 to September 2005 in Egypt by *Kifaya* (Enough!) were in fact possible thanks to *Kifaya* activists' strategy to self-limit their mobilization. *Kifaya* was politically active thanks to the use of moderated repertoires of action, a result of the limits imposed on the number of people participating in the organized demonstrations and on the choice of location of mobilization. *Kifaya* paid careful attention to the extent of mobilization, never exceeding a thousand people, and its location, mobilizing in downtown Cairo rather than in densely populated areas where too many people could gather. This enabled the network to repeatedly denounce domestic issues related, for instance, to President Hosni Mubarak's repressive regime and his attempts to enact hereditary succession (Duboc 2011: 61; Vairel 2011: 32; Beinin 2011: 185).

In addition to the *Kifaya* network, other organizations prepared the terrain for the protests observed in 2011. Some Muslim Brotherhood members from Alexandria and supporters of the Revolutionary Socialists, after years of debate over the correct form of organizational structure to follow, formed the National Alliance for Change and Unions within universities in 2005 (Manduchi 2014). Together with *Kifaya* activists, the Revolutionary Socialists were among those who took part in the anti-police riots that broke out after the murder of the young activist Khaled Said in Alexandria in 2010 by a police officer. Moreover, women played a key role in mobilizing dissent, as happened thanks to the development of a significant grassroots women's movement prior to 2011, for example, within NGOs like the Alliance for Arab Women (Amar and Lababidy 1999). Thus, women have been protagonists of the 2011 street protests, and have been relevant within both workers' movements and Popular Committees (Biagini 2020), despite the fact that they have been frequently attacked by ruling governments (Pilati et al. 2019).

Studies have also shown the role of social Islamist organizations. As Wickham (2002) explains, Islamist groups had more success overcoming authoritarian constraints than their secular rivals did. In Egypt, new opportunities for Islamic organizations and outreach began to emerge on the periphery. In other words, social, cultural, and economic groups and networks enabled citizens to participate in public life but did not compete for political power (Wickham 2002: 13).

By promoting new values, identities, and commitments, the Islamists had created new motivations for action. For instance, the graduates' embrace of an ideology was based on framing activism as a "moral obligation" (Wickham 2002: 148–151). Islamist outreach to educated youths took place in local mosques, community associations, informal study groups, summer camps, and peer networks, the building blocks of a vast, decentralized Islamic sector with substantial autonomy from state control (Wickham 2002: 16).

Below we investigate whether and how some of the aforementioned groups, despite different organizational forms, were crucial for the development of such events.

The empirical study

Data sources

The fieldwork research comprised 58 semi-structured interviews undertaken by the first author (Acconcia 2018).[4] Twenty semi-structured interviews were conducted with male and female, Islamist and Secular Egyptian activists: Revolutionary Socialists, *Kifaya*, Socialist Alliance (16 percent), and Young Islamists (18 percent); nine with male (7 percent) and female (8.5 percent) trade unionists; 29 with male (27 percent) and female (23.5 percent) workers; and ordinary citizens involved in grassroots mobilizations (e.g., Popular Committees) in Cairo and Mahalla al-Kubra between 2011 and 2015.

The empirical research involved Egyptian activists with low- and high-education backgrounds, middle- and upper-middle class citizens (44.5 percent), working-class (15.5 percent), and lower-middle class Egyptians (40 percent). Some of the interviews were conducted through a number of collective discussions. The testimonies offered insights and perspectives of the post 2011 uprisings in urban and peripheral Egyptian neighborhoods.

As for the interviewees involved in Popular Committees, after a first meeting with an ECESR gatekeeper (Centre for Economic and Social Rights), a snowball method was utilized to involve other participants. Thus, the selection of the interviewees was based on contacts from initial members active in the local committees to additional participants via chain referral in order to select both civil society activists and ordinary citizens.

In addition, the gatekeepers working as NGO activists and trade unionists were interviewed in Cairo. They formed part of the process for the composition and organization of the interviews carried out in Mahalla al-Kubra. The semi-structured interviews were organized with the specific aim of understanding: the workings of grassroots mobilization and police repression; levels of mobilization within the social movements; cooperation between the oppositional groups; personal changes in political participation of specific activists after the 2013 military coup; narratives of the 2011 uprisings and their aftermath; relations with state agencies, political parties, and the Muslim Brotherhood, and targets and strategies of these organizations. At the end of each meeting we had a debriefing session with the gatekeepers involved in order to talk about the group dynamics and the relevant results for their activities.

Those among the interviewees who were supporters of the Muslim Brotherhood appeared to be supportive of the changing nature of the activities of the group from a primarily apolitical service organization, prior to February 2011, to a political party.[5] This meant a more open participation to public protests between January 2011 and June 2011 (especially on Fridays between Tahrir Square protests and the Agouza protests), on the eve of the Mohammed Mahmoud Street clashes, in November 2011, and in June 2012 and June 2013, supporting the legitimacy of the elected president, Mohammed Morsi.[6] Those among the unionized workers and farmers interviewed in Mahalla al-Kubra had been part of several waves of protests before the 2011 uprisings, especially within *Kifaya* and April 6 Youth Movements. However, this participation in many cases hadn't been clearly formalized and remained at the individual level, although some of the interviewees had been affiliated to a trade union since the 80s.[7]

Access to the field was very problematic, especially as a consequence of the increasingly repressive measures taken after the 2013 military coup in Egypt. At the beginning, the interviewees did not express any security concerns with reference to their participation in the interviews. However, after the 2014 presidential elections in Egypt, the local trade unionists involved in the interviews conducted in Mahalla al-Kubra appeared to be more concerned about voicing their opinions.[8] Some of the workers asked to be mentioned only with their first names in order to be less noticeable. As a consequence, all interviewees have been anonymized and each interviewee was assigned a number.

Variety of groups and the 2011 protests in Egypt

The Muslim Brotherhood

The monopolization of political dissent by Islamist groups is a common feature of many Arab and Middle Eastern countries. In *Weapons of the Weak*, James Scott (1990) explained how Islamists monopolized the space of dissent in the village of Sadaka. As Bayat (2010) noted, Scott's ethnographic studies focusing on individual reactions of peasants, along with Foucault's decentrated notion of power and the revival of the concept of Neo-Gramscian hegemony, can serve to enhance a "micro-politics" perspective on social movements. Ever since its foundation, the Muslim Brotherhood indeed operated as a substitute for the State among the lower social strata, therefore challenging the legitimacy of the ruling elite (Mitchell 1969: 169). Placing these approaches in the context of the Egyptian protests between 2010 and 2012, not only did the Islamists monopolize the opposition movements in the prerevolutionary phase but, during the uprisings, they manipulated street movements and less organized entities in order to use and then deactivate their revolutionary potential. At the very beginning of the occupation of Tahrir Square on January 25, 2011, in a wave of high political mobilization and solidarity between the movements, there was "noncompetitive cooperation" (Della Porta & Diani 2006: 157) between the different groups. This was helped by the permanent occupation of the same public spaces. After the dismissal of Hosni Mubarak on February 11, 2011, the Muslim Brotherhood and other opposition groups witnessed, in contrast, a "competitive cooperation" (ibidem). This phase lasted until the Mohamed Mahmud Street clashes in November 2011. At this stage, despite a long internal debate about the need to forge a political party, different waves of state repression and increased engagement in grassroots associations, the

Muslim Brotherhood decided to formalize their political party Freedom and Justice (FJP) and took part in the electoral process (Ketchley 2017). This had consequences on their members' engagement in protests. As a female activist who took part in Tahrir Square demonstrations stated that:

> *During the days of fights on Mohammed Mahmud Street, the Muslim Brotherhood abandoned the youth of Tahrir in the streets.*[9]

The electoral victories of FJP in 2012, helped by the absence of politicians belonging to the NDP at the parliamentary elections due to their temporary ban from party politics, saw "neutrality" (Della Porta & Diani 2006: 157) prevailing among the opposition movements or newly formed political parties. This stage was backed by a wave of demobilizing political engagement and strengthened ideological sentiment of belonging. The army's stigmatization of the Muslim Brotherhood as counterrevolutionaries stimulated renewed protests that brought about a complete fragmentation of the coalition of forces of the 2011 uprisings in the wake of the July 3, 2013, military coup (Barrie & Ketchley 2018).

> We [the Muslim Brotherhood] *tried to include other opposition forces within the Constituent Assembly. That year they* [other opposition groups] *were called hundreds of times to give them responsibilities within the government. They always refused,*[10]

a male activist who took part in Tahrir Square demonstrations explained. In addition to strengthening electoral politics and demobilizing street politics, the role of the Muslim Brotherhood was crucial for the emergence, in 2011, of new means of popular mobilization, triggered by participation of many of its members in alternative networks, which included local Popular Committees.

Popular Committees

The 2011–2013 mass riots were paramount in the formation of new means of popular mobilization such as Popular Committees that aimed at enhancing a diverse range of unmet needs and motivating ordinary citizens to participate in a series of activities. These included providing social services, security and self-defense, delivering gas tanks for cooking and heating, supplying food at low prices, planning sewage systems, and bringing electricity to residents, as well as participating in the political arena (Hassan 2015). Members of Popular Committees were often supporters of the Muslim Brotherhood and of *Salafi* groups and exhibited "important continuities with Islamist activism." As we have discussed, Islamic activism includes major social charities in Egypt (El-Meehy 2012) and the latter are considered free spaces in repressive contexts. Popular Committees in Egypt were frequently rooted within the preexisting networks of Muslim Brotherhood charities, PVO, schools, and hospitals.

> *The Popular Committees have been put in place thanks to the organizational structure of the Muslim Brotherhood, their specific knowledge of the district and their capacity to identify any minimum risk. Their representatives within the charities were very useful in order to unify and manage the people taking part within the Committees,*

a male participant within Popular Committees in Cairo explained.[11] In this framework, Popular Committees played a crucial role in promoting individuals' active engagement in politics, both institutional politics and protests (El-Meehy 2012). According to the Egyptian Life for Development Foundation (El-Meehy 2012), thousands of Popular Committees (*lijan sha'biyya* in Arabic) were active in Cairo during the 18 days of occupation of Tahrir Square. In three days, between the first demonstration in Tahrir Square in Cairo on January 25, 2011, and the "Friday of Anger" on January 28, 2011, the police began to retreat or apparently disappear from the Egyptian streets and in a few hours, Popular Committees were quickly organized. "Neighborhood watch brigades, typically led by young men, sprang up to fill the security void as reports of criminal violence mounted."[12] During our interviews we talked to male and female participants in one of Berqet Fil's Popular Committees in Sayeda Zeinab about the reasons why they initially mobilized. As an interviewee stated, ordinary people were heavily involved in self-defense groups. "I spent my all day and night taking care of the safety of my neighborhood."[13] Another interviewee added that their mobilization was a direct consequence of the absence of policemen. "With the honest people of my area we formed groups to substitute the absence of policemen after their disappearance."[14] According to another interviewee:

> *The police force disappeared from the street because it was not trained to resist for days of confrontations at the micro level with the people. It has been a structural failure, caused by the interruption of communications (often brought about by a lack of a battery in their walkie-talkies).*[15]

The mobilization of the Popular Committees was a first reaction to the arbitrary methods of the police:

> *During my night shift, I often encountered former and violent policemen engaged in indiscriminate lootings.*[16]

Another interviewee added that his participation in the Popular Committees was necessary to protect his home from the spreading presence of criminals.[17]

According to El-Meehy (2012), in some districts the Committees continued to gather in spring and summer of 2011 to discuss the main problems of the neighborhood: "cleaning streets, fixing water fountains to improve living conditions in the area and painting buildings." Furthermore, in the neighborhood of Basatin in the Cairo Governorate where she focused her research, the members of Popular Committees "gradually turned their attention to politics," evolving toward "active citizenship" (El-Meehy 2012).

The Committee's participants were also involved in the electoral campaign for the constitutional amendments in the March 2011 referendum, although many participants had returned to their daily life and shared some of the mainstream opinion, which stigmatized the remaining activists, who were pictured as a source of instability and therefore against Egyptian national interests. In 2011 and 2012, many of the interviewees were also engaged in the more institutional pursuit of electoral campaigns and in party politics.

The majority of the interviewees supported the Muslim Brotherhood at the ballot boxes during the November–January 2011–2012 parliamentary elections.

> *I was interested in Freedom and Justice Party (FJP). Thus, I decided to vote for them at the parliamentary elections,*[18]

interviewed male and female participants within Popular Committees in Cairo stated. They argued that the Muslim Brotherhood supporters encouraged their constituency to participate in the electoral process promising different kind of rewards.

> *The supporters of the Muslim Brotherhood and Salafi groups, previously present within the Popular Committees, were distributing food, sugar, oil and clothes (galabyyas) at the school entrances to encourage their supporters to vote for them,*[19]

he argued. In this framework, the participants in the Popular Committees, especially if young, students, or unemployed, had been the first to be ready to take part during the continual waves of electoral mobilization and campaigns. Some of them appeared to be motivated by more conscious revolutionary and secular intentions:

> *We wanted a new Constitution. For this reason, we distributed flyers asking to the people to vote No.*[20]

Once again, they were confronted with an electoral choice on the occasion of the 2012 presidential elections.[21] They decided to vote for Mohammed

Morsi only to prevent the election of the former Mubarak regime prime minister, Ahmed Shafiq.

> *We were not happy with the Muslim Brotherhood but we did not want a felul (man of the old regime) to be the new president.*[22]
>
> *Some of the members of our Committee during the days of the revolution encouraged people of my building to go to vote for the Brotherhood representative. Many of them did it for the relationships of trust built-up especially during the previous months of mobilization.*[23]

On the other hand, young participants within Popular Committees began their boycotts of the electoral process.

> *Leftist parties were not ready to prepare a campaign. I could never expect that a politician coming from the Muslim Brotherhood could have been chosen as the new Egyptian president.*[24]

According to El-Meehy (2012), the Popular Committees were successfully engaged in more ambitious projects as well.

> *Ard al-Lewa's Committee self-financed a railway crossing to minimize accidents among residents. It also mobilized around the establishment of a park, school and a hospital on fourteen feddans of vacant land owned by the Ministry of Religious Endowments (Awqaf) in the neighborhood. Next door, the committee in Imbaba organized effective non-payment campaigns for public services the state failed to provide, such as garbage collection, while Nahia's Committee constructed an on/off ramp to connect the neighborhood to the ring road.*[25]

By doing so the Popular Committees were slowly becoming NGOs, tending to merge with the preexisting networks of Muslim Brotherhood charities, schools, and hospitals.

As an interviewee confirmed,

> *In Berqet Fil, many participants within the Popular Committees were involved in associations working with the elders or providing social services to the disabled.*[26]

Other interviewees from the Popular Committees began working in centers and NGOs focused on the defense of human rights.

> *My participation in the grassroots movements has been vital for my present work position as a human rights defender*[27]

he added. However, after the 2013 military coup in Egypt, all the charities of the Muslim Brotherhood, its hospitals, its NGOs, its associations, and its media outlets either were closed down or faced noticeable levels of repression or the removal of their former management. The Society of the Muslim Brotherhood, its political party the FJP, and the coalition defending the Morsi government's legitimacy were all outlawed by the Egyptian courts. Finally, the Islamist movements within the universities were heavily repressed (especially on the Al-Azhar and Ayn Shamps campuses). After the Rabaa massacre (August 14, 2013), interviewed male and female participants within Popular Committees did not take any further part in the electoral processes or in demonstrations.

> We boycotted the Constitutional Referendum (January 2014), presidential and parliamentary elections (May 2014, December 2015),[28]

participants within Popular Committees in Cairo stated.

Trade unions and workers' mobilization

Trade unions did not act as structures of resource mobilization for the 2011 protests. First and foremost, ETUF, the official trade union, did not support protests, in continuity with its behavior in previous years. Controlled by the Mubarak regime, ETUF did not call for labor protests during the 2011 uprisings, even if many groups were clearly ready for a mass mobilization. The spontaneous workers' committees acting at the local level had no "institutional mechanism to compel the ETUF to join the popular movement against Mubarak" (Beinin 2016: 107). In other words, the local committees were not duplicated at the national level through an organizational structure capable of coordinating the local level actions. Many workers were initially organized in more spontaneous oppositional mobilization. As argued by Beinin (2011: 183), even workers' protests in Egypt between 2006 and 2009 did not rely on "movement entrepreneurs" or preexisting organizations.

With the exception of the support from several labor-oriented NGOs, workers' protests in Egypt mainly relied on occasional face-to-face meetings and mobile telephones, supported by family and neighborhood connections (Beinin 2011: 183). The working-class networks were thus highly localized, whereby family and neighborhood connections were of utmost importance in the daily life and in the construction of workers' neighborhoods. Spontaneous workers' groups found an institutional form only after the protests broke out. Indeed, on January 30, 2011, the CTUWS coordinator, Kamal Abbas, and the RETAU president, Kamal Abu Eita, along with smaller unions of teachers, health professionals, and retiree associations formed the EFITU. This occurred despite the fact that ETUF continued to support the state institutions while in April 2011 its president, Hussein Megawer, was arrested for his affiliation to the then dissolved NDP and the individual presence of Muslim Brotherhood figures within its members was growing, especially during their year in power (2012–2013).[29]

The Tahrir Square demonstrations encouraged the workers' groups to mobilize, communicate, and build intergroup networks (Tripp 2013: 160). In early 2011, the nationwide teachers' strike involving half a million workers demanded the *cleansing* (*tathir*) of public institutions of the remnants of the old regime (Hanieh 2013: 169). In February 2011, 489 strikes occurred in Egypt. The EFITU issued a statement proclaiming "Demands of the Workers in the Revolution": the right to form non-governmental unions, the right to strike and the dissolution of the pro-regime and corrupt ETUF. The SCAF (Supreme Council of the Armed Forces) appointed Ahmed el-Borai, professor of labor law at Cairo University, as the interim Minister of Manpower. Therefore, labor mobilizations were constantly increasing in parallel with some trade unionists' attempts to have better represented workers' rights within the interim government.

In April 2013, a second federation of independent unions was established: the EDLC convened with 149 unions represented.[30] Mobilizations continued until the military junta intervened to put all kind of protests under state control. Thus, while El-Borai promoted a draft law of EFITU in August 2011, a law criminalizing strikes, demonstrations and sit-ins was also approved (Tripp 2013: 161; Abdelrahman 2015).

Evidence in our qualitative research confirmed workers' early participation in the 2011 protests in Mahalla al-Kubra even if the number of participants was not comparable to the Tahrir Square mass riots. Moreover, many of them had already been involved in other previous anti-regime mobilizations and strikes. As mentioned, according to our interviewees the workers' movement had been spontaneously activated,[31] but workers had prior individual experience in SMOs and protests. Among the protesters there were many long-term supporters of anti-Mubarak movements, especially *Kifaya*: "We were previously involved in the *Kifaya* movement," a male unionized worker stated,[32] while others had already taken part in the 2006–2008 labor strikes. "We participated during the first protests after an already long-lasting struggle to overcome the rooted crisis of the Egyptian cotton industry,"[33] male and female unionized workers in Mahalla al-Kubra added. Other interviewees continued:

> *There were many contradictions in the working class. We worked to bring the factories to Tahrir and vice versa. Our slogan was the Square and the factory one hand.*[34]

As for the period of political transition, initially, for many workers it seemed wise to take part in the 2011–2012 parliamentary elections although many other workers – in Cairo, Suez, and Alexandria – were already feeling marginalized within the political arena and decided to boycott the elections from that stage onwards.

> *I decided to go to vote and support independent workers or some candidates of the local party al-Adl (Justice),*[35]

a female unionized worker stated. On the occasion of the 2012 Constitutional Referendum, the Mahalla al-Kubra workers and farmers were already very critical of the political approach of the Muslim Brotherhood. The "No" to the new Constitution here won with 52%. In more general terms, all the policies implemented at the national level by the Islamists appeared to be ineffective in supporting workers' rights. "The 2012 Constitution was against workers' rights,"[36] an RS female activist added.[37] In 2015 a number of new strikes in textile, cement, and building factories began. According to our interviewees, those strikes were neither structured nor well organized.

> *The fear among workers, farmers and all other opposition groups is unprecedented within the framework of al-Sisi's military regime.*[38]

highlighted an interviewee.

> *The December 2015 meeting at the Center for Trade Union and Workers Services (CTUWS) in Cairo was especially important because it was a first attempt to coordinate again the works of the local fragmented and isolated unions. On this occasion we decided to forge a committee representing workers' rights and to launch a national campaign for supporting trade union freedoms,*[39]

a female unionized worker in Mahalla al-Kubra added.

Discussion and conclusion

The chapter aimed to understand the role of different types of mobilizing groups during and in the aftermath of the protests that started in 2011 in Egypt. It tried to encompass the analysis of both loose and informal groups, more structured organizations and their interconnections. Empirically, the study drew on a qualitative analysis, using data collected through semi-structured interviews undertaken between 2011 and 2015 to members of Popular Committees (PC), unionized workers, members of the Revolutionary Socialists (RS), and *Kifaya* movement. Our findings show that the protests in 2011 were not fully spontaneous and that a preexisting organizational structure was at work prior to the eruption of the protests. While a widespread literature has highlighted the spontaneous nature of the 2011 protests in the MENA region, this study also highlights the crucial role of coordination that groups with different degrees of formalization did play, often in interconnection with each other.

Results show that while several organizations were not actively engaged in political activities, thus proving ineffective for their members' political participation in protests, as in the case of ETUF, other organizations like charities and apolitical organizations did provide the structural basis and those free spaces for people to engage politically.

This occurred, for instance, thanks to the links that established organizations such as the Muslim Brotherhood provided with more informal groups, like the Popular Committees. Likewise, more spontaneous and loosely structured groups equally provided those resources, links, and ties for more structured and established organizations to emerge, as occurred for self-organized workers' groups, which eventually institutionalized into EFITU.

This often occurred thanks to the link that established organizations such as the Muslim Brotherhood provided with more informal groups, like the Popular Committees. In turn, more spontaneous and loosely structured groups equally provided those resources, links, and ties for more structured and established organizations to emerge, as occurred for self-organized workers' groups which eventually institutionalized into EFITU.

Consequently, loosely organized groups were closely linked with more coordinated or established organizations, groups renovated their forms, and individuals mixed formal membership with informal ties, and were ready to pass from loosely structured groups into established organizations and vice versa. Several members of Muslim Brotherhood and *Salafi* groups eventually engaged in political forms of actions thanks to individual involvement of their members in more loosely structured forms like Popular Committees. Likewise, but in the opposite direction, when protests broke out in January 2011 self-organized workers' groups converged into an established coordinating body through the legalization of EFITU. Evidence of similar processes has been found in other repressive contexts. The 2013 Gezi Park protests, the largest civilian uprisings in the last decade in Turkey, which began in late May and lasted until September 2013, was initially a spontaneous revolt. However, Gezi Park became a focal point for a larger movement composed of a diversity of groups and organizations such as, inter alia, workers, unions, student organizations, and non-Turkish organizations (Anisin 2016: 415). More importantly, "new activist groups emerged and stemmed from the original Taksim Solidarity movement such as Occupy Gezi Park" (Anisin 2016: 423). Likewise, the 2016 mass protests in Poland against a bill that would impose criminal sanctions on abortion were largely mobilized in online and informal spaces, but they also succeeded in getting women out on the streets through the presence of more formal pro-choice groups (cf. Soon & Cho 2014 for a similar dialectic between different spaces of mobilization in Singapore).[40]

In light of this evidence, future research may strengthen the comparative dimension of the results examining protests in neighboring countries. In addition, while our findings clarify the importance of considering the variety of groups, research may investigate further how groups with different organizational forms can eventually support different type of actions. The type and scope of collective actions that more informal, small, and loosely structured groups can engage in are likely to differ from the type of collective actions promoted by organizations. This had been already discussed by

Melucci (1996), who drew attention to the presence and activities of informal groups in the so-called new social movements of the 1980s.

As studied by Melucci, informal groups acting together tend to form a segmented, reticular and multifaceted, often loose, network structure. Due to these characteristics, they can promote collective actions. However, the latter risk being too often narrow in scope and grounded at the local level. The networks built by the new social movements indeed profoundly differed from the image of the networks formed by politically organized actors.

The strength of informal groups and networks lies precisely in their provision of flexibility, adaptability, and immediacy, which more structured organizations cannot incorporate (Melucci 1985: 800, 1996: 115). While these characteristics are of utmost importance for actors in repressive contexts, they can also become weaknesses to the degree that collective actions promoted by such groups cannot be coordinated on a large scale, neither can they be cumulative or sustained in time, an aspect typical of dynamics of social movements (cf. Tilly 1978).

Indeed, in the Egyptian case, only after the institutionalization of workers' spontaneous groups through the establishment of the EFITU, were workers able to coordinate national-level strikes throughout the first half of 2011 and later on, far beyond parochial and local actions. More specifically, only after the set-up of EFITU and the approval of the law on the legalization of independent trade unions, from April to September 2011, was there a rapid expansion of labor organizations and a spread of independent unions in Mahalla al-Kubra.

The strikes of September 2011 rather than occurring as isolated and in fragmented workplaces were indeed supported by 500,000 workers nationwide (Alexander & Bassiouny 2014: 213).

Notes

1 This chapter has been previously published in a different format in International Sociology Acconcia, G., & Pilati, K. (2020). Variety of groups and protests in repressive contexts: the 2011 Egyptian uprisings and their aftermaths. *International Sociology, 36*, 91–110.

2 Tilly (1978: 62–63) discussed this mechanism through the concept of CATNET, a synthesis of *catness* and *netness*. While the former, *catness*, identifies the presence of an aggregate of individuals defined by specific external categorical traits (young people, women, immigrants), the latter, *netness*, identifies the presence of stable relationships. Through the concept of CATNET Tilly discusses the relationships which facilitate, if shared categorical traits among actors exist, the passage from a social category to a social group capable of acting intentionally. Thanks to the presence of intense and durable relations between individuals with common external categorical traits, social categories can transform into social groups whose members share interests and identities on the basis of which can act collectively. Therefore, collective actions depend on the level of *catnet*, which

refers both to the characteristics linked to a certain social category and to the density of networks created by individuals who share those social traits.

3 Not all authors agree on the mobilizing role of online ties (Brym et al. 2014). Hassanpour (2014) shows that protest increased in the period when the internet was switched off, January 28, 2011, while Clarke (2014) argues that social media helped during the first day but had a negligible impact thereafter.

4 These interviews are part of a large fieldwork research which comprised several methodological techniques, including focus groups and field work notes. We chose to rely on the material drawn from the interviews because it provided the richest information for the specific aims of this study.

5 Interviews 7, 8, and 12, Cairo.

6 Interviews 14, 15, Mahalla al-Kubra.

7 The repression strongly affected the workings of independent trade unions that have been officially banned in 2017. See https://carnegieendowment.org/sada/64634 [Last accessed July 28, 2020].

8 Interviewee 42, Cairo.

9 Interviewee 35, Cairo.

10 Interviewee 7, Cairo.

11 Interviewee 2, Cairo.

12 Interviewee 1, Cairo.

13 Interviewee 4, Cairo.

14 Interviewee 1, Cairo.

15 Interviewee 3, Cairo.

16 Interviewee 5, Cairo.

17 Interviewees 7 and 8, Cairo.

18 Interviewee 3, Cairo.

19 Interviewees 1, 2, 3, 6, and 7, Cairo.

20 Interviewees 4 and 5, Cairo.

21 Interviewee 2, Cairo.

22 Interviewee 3, Cairo.

23 Interviewees 4 and 5, Cairo.

24 Interviewees 4 and 5, Cairo.

25 Interviewee 7, Cairo.

26 Interviewee 4, Cairo.

27 Interviewees 1 to 8, Cairo.

28 Interviewees 1 to 8, Cairo.

29 https://carnegieendowment.org/sada/50540 [Last accessed July 31, 2020].

30 www.madamasr.com/sections/politics/whatever-happened-egypts-independent-unions [Last accessed May 10, 2016].

31 Interviewees 15, 18, and 17, Mahalla al-Kubra.

32 Interviewee 19, Mahalla al-Kubra.

33 Interviewees 20, 21 and 22, Mahalla al-Kubra.

34 Interviewee 18, Mahalla al-Kubra.

35 Interviewee 14, Mahalla al-Kubra.

36 Interviewee 16, Mahalla al-Kubra.

37 Popular Committees, workers' groups, members of organizations such as *Kifaya* or the Revolutionary Socialists were not the only groups active in the 2011 protests. Evidence also shows that on January 25, 2011, the starting day of the occupation of Tahrir Square in Cairo, a march of 10,000 people was led by the leader of the al-Ahly football fan club league in Cairo and during the 18-day occupation of Tahrir Square, the ultras also patrolled the perimeters of the square and controlled entry (Dorsey 2012: 414).

38 Interviewee 22, Mahalla al-Kubra.

39 Interviewee 22, Mahalla al-Kubra.
40 Nawrkoicz Kasia Czarny Protest: how Polish women took to the streets www.opendemocracy.net/en/can-europe-make-it/czarny-protest-how-polish-women-took-to-streets/

References

Abdelrahman, M. (2015). *Egypt's long revolution protest movements and uprisings.* New York, NY: Routledge.

Acconcia, G. (2018). The uprisings in Egypt: Popular committees and independent trade unions (PhD thesis). London: Goldsmiths College, University of London.

Achcar, G. (2013). *The people want. A radical exploration of the Arab uprising.* London: Saqi Books.

Alexander, A., & Bassiouny, M. (2014). *Bread and freedom, social justice. Workers and the revolution.* London: Zed Books.

Amar, N. H., & Lababidy, L. S. (1999). Women's grassroots movements and democratization in Egypt. In J. M. Bystydzienski & J. Sekhon (Eds.), *Democratization and women's grassroots movements* (pp. 150–170). Bloomington, IN: Indiana University Press.

Anisin, A. (2016). Repression, spontaneity, and collective action: The 2013 Turkish Gezi protests. *Journal of Civil Society, 12*(4), 411–429.

Barrie, C., & Ketchley, N. (2018). Opportunity without organization: Labor mobilization in Egypt after the 25th January revolution. *Mobilization: An International Quarterly, 23*(2), 181–202.

Bayat, A. (2010). *Life as politics: How ordinary people change the Middle East.* Amsterdam: Amsterdam University Press.

Beinin, J. (2011). A workers' social movement on the margins of the global neoliberal order, Egypt 2004–2009. In J. Beinin & F. Vairel (Eds.), *Social movements, mobilization and contestation in the Middle East and North Africa* (pp. 181–201). Stanford, CA: Stanford University Press.

Beinin, J. (2016). *Workers and thieves, labor movements and popular uprisings in Tunisia and Egypt.* Stanford, CA: Stanford University Press.

Biagini, E. (2020). Islamist women's feminist subjectivities in (r)evolution: The Egyptian Muslim sisterhood in the aftermath of the Arab uprisings. *International Feminist Journal of Politics, 22*(3), 382–402.

Bremer, J. A. (2012). Leadership and collective action in Egypt's popular committees: Emergence of authentic civic activism in the absence of the state. *The International Journal of Not-for-Profit Law, 13*(4), 70–92.

Brym, R., Godbout, M., Hoffbauer, A., Menard, G., & Zhang, T. H. (2014). Social media in the 2011 Egyptian uprising. *The British Journal of Sociology, 65*(2), 266–292.

Clark, J. A. (2004a). *Islam, charity, and activism – Middle-class networks and social welfare in Egypt, Jordan, and Yemen.* Bloomington, IN: Indiana University Press.

Clarke, K. (2014). Unexpected brokers of mobilization: Contingency and networks in the 2011 Egyptian uprising. *Comparative Politics, 46*(4), 379–397.

Della Porta, D. (2014). *Mobilizing for democracy, comparing 1989 and 2011.* Oxford: Oxford University Press.

Della Porta, D., & Diani, M. (2006). *Social movements: An introduction*. Oxford: Blackwell Publishing.

Dorsey, J. M. (2012). Pitched battles: The role of ultra soccer fans in the Arab Spring. *Mobilization: An International Journal, 17*(4), 411–418.

Duboc, M. (2011). Egyptian leftist intellectuals' activism from the margins: Overcoming the mobilization/demobilization dichotomy. In J. Beinin & F. Vairel (Eds.), *Social movements, mobilization, and contestation in the Middle East and North Africa* (pp. 61–79). Stanford, CA: Stanford University Press.

El-Meehy, A. (2012). Egypt's popular committees from moments of madness to NGO dilemmas. *Middle East Research and Information Project, 42*(265). Retrieved March, 2020, from https://merip.org/2013/01/egypts-popular-committees/

Goldstone, J. A. (2016). Regimes, resources, and regional intervention – Understanding the openings and trajectories for contention in the Middle East and North Africa. In Y. A. Eitan, S. Avraham, & S. Mario (Eds.), *Popular contention, regime, and transition: Arab revolts in comparative global perspective* (pp. 97–114). Oxford: Oxford University Press.

Gould, R. (1991). Multiple networks and mobilization in the Paris Commune, 1871. *American Sociological Review, 56*(6), 716–729.

Hamanaka, S. (2020). The role of digital media in the 2011 Egyptian revolution. *Democratization, 27*(5), 777–796.

Hanieh, A. (2013). *Lineages of revolt, issues of contemporary capitalism in the Middle East*. Chicago, IL: Haymarket Books.

Hassan, H. (2015). Extraordinary politics of ordinary people: Explaining the micro dynamics of popular committees in revolutionary Cairo. *International Sociology, 30*(4), 383–400.

Hassanpour, N. (2014). Media disruption and revolutionary unrest: Evidence from Mubarak's Quasi-experiment. *Political Communication, 31*(1), 1–24.

Hinnebusch, R. (2015). Introduction: Understanding the consequences of the Arab uprisings – Starting points and divergent trajectories. *Democratization, 22*(2), 205–217.

Howard, P. N., & Hussain, M. M. (2013). *Democracy's fourth wave?: Digital media and the Arab spring*. New York, NY: Oxford University Press.

Jamal, A. (2007). *Barriers to democracy: The other side of social capital in Palestine and the Arab world*. Princeton, NJ: Princeton University Press.

Ketchley, N. (2017). *Egypt in a time of revolution: Contentious politics and the Arab spring*. New York, NY: Cambridge University Press.

Klofstad, C. C. (2011). *Civic talk: Peers, politics, and the future of democracy*. Philadelphia, PA: Temple University Press.

Lichterman, P. (2008). Religion and the construction of civic identity. *American Sociological Review, 73*(1), 83–104.

Lim, M. (2012). Clicks, cabs, and coffee houses: Social media and oppositional movement in Egypt. *Journal of Communication, 62*(2), 231–248.

Manduchi, P. (2014). *I movimenti giovanili nel mondo arabo mediterraneo*. Roma: Carocci editore.

McAdam, D., McCarthy, J., & Zald, M. (Eds.). (1996). *Comparative perspectives on social movements*. Cambridge Studies in Comparative Movements. Cambridge: Cambridge University Press.

McCarthy, J. D., & Zald, M. N. (1977). Resource mobilization and social movements: A partial theory. *American Journal of Sociology, 82*(6), 1212–1241.

McVeigh, R., & Sikkink, D. (2001). God, politics, and protest: Religious beliefs and the legitimization of contentious tactics. *Social Forces, 4,* 1425–1458.

Melucci, A. (1985). The symbolic challenge of contemporary movements. *Social Research, 52*(4), 789–816.

Melucci, A. (1996). *Challenging codes: Collective action in the information age.* Cambridge: Cambridge University Press.

Mitchell, R. P. (1969). *The society of the Muslim brothers.* Oxford: Oxford University Press.

Pfaff, S. (1996). Collective identity and informal groups in revolutionary mobilization: East Germany in 1989. *Social Forces, 75*(1), 91–117.

Pilati, K., Acconcia, G., Suber, L., & Chennaoui, H. (2019). Between organization and spontaneity of protests: The 2010–2011 Tunisian and Egyptian uprisings. *Social Movement Studies, 18*(4), 463–481.

Scott, J. C. (1990). *Weapons of the weak: Everyday forms of peasant resistance.* Oxford: Oxford University Press.

Soon, C., & Cho, H. (2014). OMGs! Offline-based movement organizations, online-based movement organizations and network mobilization: A case study of political bloggers in Singapore. *Information, Communication & Society, 17*(5), 537–559.

Steinert-Threlkeld, Z. C. (2017). Spontaneous collective action: Peripheral mobilization during the Arab spring. *American Political Science Review, 111*(2), 379–403.

Tarrow, S. (1996). States and opportunities: The political structuring of social movements. In D. McAdam, J. McCarthy, & M. Zald (Eds.), *Comparative perspectives on social movements. Cambridge studies in comparative movements* (pp. 41–61). Cambridge: Cambridge University Press.

Tilly, C. (1978). *From mobilization to revolution.* Boston, MA: Addison-Wesley.

Trejo, G. (2012). *Popular movements in autocracies: Religion, repression, and indigenous collective action in Mexico.* Cambridge: Cambridge University Press.

Tripp, C. (2013). *The power and the people, paths of resistance in the Middle East.* Cambridge: Cambridge University Press.

Vairel, F. (2011). Protesting in authoritarian situations: Egypt and Morocco in comparative perspective. In J. Beinin & F. Vairel (Eds.), *Social movements, mobilization, and contestation in the Middle East and North Africa.* Stanford, CA: Stanford University Press.

Verba, S., Schlozman, K. L., & Brady, H. E. (1995). *Voice and equality: Civic voluntarism in American politics.* Cambridge, MA: Harvard University Press.

Wickham, C. R. (2002). *Mobilizing Islam: Religion, activism, and political change in Egypt.* New York, NY: Columbia University Press.

3 Protest demobilization in postrevolutionary settings[1]

with Katia Pilati

This chapter examines two outcomes of demobilization in postrevolutionary contexts, democratic transition and counterrevolution. Complementing elite-driven approaches, we argue that the way demobilization ends is conditional upon the capacity of challengers to promote enduring alliances. Following a paired controlled comparison, we analyze two cases, Egypt and Tunisia and processes of alliance building and fragmentation preceding the 2013 *coup d'Etat* in Egypt, and the adoption of a new Constitution in 2014 in Tunisia. Data from semi-structured and in-depth interviews were collected through fieldwork in multiple localities of Egypt and Tunisia between 2011 and 2019. Results show that the fragmentation of the challengers' coalition in postrevolutionary Egypt contributed to a counterrevolution while, in Tunisia, challengers' alliances rooted in the prerevolutionary period lasted throughout the phase of demobilization and supported a democratic transition. We conclude by discussing some alliance-based mechanisms accounting for a democratic transition: intergroup trust-building, brokerage, and ideological boundary deactivation.

Introduction

In this chapter, we focus on the process of demobilization and two opposed outcomes of such a phase that may occur in post-authoritarian contexts: governments' democratic transition and governments' counterrevolution. We examine the process of demobilization that followed the peak of the 2010–2011 protests in Egypt and Tunisia and in particular, we investigate the role of challengers' networks in shaping the two different outcomes. We examine bottom-up processes that may integrate existing accounts on democratization and counterrevolutions focused on elite-driven approaches. We share Geddes' (2011) argument that: "transitions from personalist dictatorships are seldom initiated by regime insiders; instead, popular opposition, strikes, and demonstrations often force dictators to consider allowing multiparty elections (Bratton & van de Walle 1997). . . . The process of transition from personalized dictatorship should not be modelled as an elite-led bargain." By drawing on the literature of contentious politics, we

DOI: 10.4324/9781003293354-4

propose two types of bottom-up mechanisms that may explain divergent outcomes of demobilization. First, we discuss broad mechanisms of alliance building. We argue that the way demobilization ends and the way interactions between authorities and challengers stabilize, within either a democratic framework or a counterrevolution, are conditional upon the capacity of challengers to promote enduring alliances throughout the phase of demobilization.

While alliances among challengers are crucial to sustain revolutionary coalitions during the phase of mobilization, their durability will support the transition to democratic rules. In contrast, fragmentation of challengers' prerevolutionary coalitions during processes of demobilization will contribute to a counterrevolution, profiting from challengers' divisions. Second, we discuss specific mechanisms contributing to explaining why alliances may favor a democratic transition. We contend that alliances imply the presence of actors bridging across different factions, high levels of trust, and ideological boundary deactivation as well as shared ideologies and frames which may all support alliance building.

Our empirical analysis focuses on the Egyptian and Tunisian case studies. We follow the comparative method of controlled comparison and select these two countries as case points of different trajectories of processes of demobilization. Despite many similarities shared by the two countries until early 2013 (Hassan et al. 2020: 555), in Tunisia a democratic transition culminated in the new Constitution, adopted on January 14, 2014. Conversely, in Egypt a military junta carried out a coup in July 2013. To understand such different outcomes, we investigate the alliances by challengers, especially during the phase of demobilization.

The latter occurred between December 2010 and January 2011 when protests in the two countries peaked, and July 2013 and January 2014 when, respectively, the *coup d'Etat* took place in Egypt and a new Constitution was adopted in Tunisia.[2] The empirical analysis uses data from semi-structured and in-depth interviews collected through several fieldworks in multiple localities of Egypt and Tunisia between 2011 and 2019. Results show that the new elites in Tunisia were able to build upon long-lasting alliances already rooted in the prerevolutionary period. Trade unions, civil society organizations, and previously banned political parties such as *Ennhada* all played a crucial role in building a comprehensive alliance that lasted throughout the postrevolutionary period. Such alliances supported the transition to democracy and the application of a new constitution in the postrevolutionary period (Durac 2019; Hassan et al. 2020; Zemni 2016). In contrast, major actors in Egypt who had loosely come together on the shared interest to oppose Mubarak, including the Muslim Brotherhood and the secular groups that were part of the revolutionary coalition that existed before 2011, ended up dividing into narrow partisanships, leading to a process of fragmentation in postrevolutionary Egypt that did not allow the formation of long-lasting alliances (Ketchley 2017: 2). This provided fertile grounds for a backlash of authoritarianism and a counterrevolution, which culminated in the 2013 coup.

Ultimately, we argue that, in addition to elite-driven dynamics of democratic transition, bottom-up processes play a fundamental role for democratic transitions, which occur when challengers' coalitions are able to endure beyond any success reached during the peak of protests.

Demobilization in postrevolutionary settings

Demobilization characterizes the decline of the protest wave, and involves a reduction of contentious action following the shrinking of the resources available to challengers for collective claims-making. Demobilization definitively ends once interactions between challengers and elites stabilize, implying more predictable routinized interactions between challengers and governments (Tilly & Tarrow 2015: 120). The decline of contention is associated with the "convergence on a new equilibrium in which neither party can hope to make substantial gains by continuing to raise the stakes of contention" (Koopmans 2004: 38).

In postrevolutionary settings, such equilibrium may result in a democratic transition. However, it may also result in a counterrevolution, defined as "collective and reactive efforts to defend the status quo and its varied range of dominant elites against a credible threat to overturn them from below" (Slater & Rush Smith 2016: 1475). Counterrevolution also brings political order with it to the degree that it aims to preserve the status quo.[3]

Scholars who have examined processes of demobilization within contentious politics have highlighted a variety of factors affecting the way demobilization ends. They are mostly focused on the changes in attitudes and actions by elites and challengers, making up the main actors in contentious dynamics.

On the elites' side, demobilization is affected by the strategies that elites use in response to challengers' claims. Elites' decisions affecting demobilization processes include the degree of reforms conceded, coupled with the level of repression used (Tarrow 1989; Della Porta 1995).

Pacts and compromise among and within elites are crucial during the processes of demobilization, positively impacting on the emergence of sustainable democracies (Kadivar 2018: 391). Considering the challengers' side, demobilization may be affected by individual exhaustion from participating in protests, including the emotional distress of protesters (Tilly & Tarrow 2015) or by challengers' resorting to violent tactics, including the use of political violence, or by their involvement in institutional actions (Tarrow 1989; Della Porta 1995).

Furthermore, the outcomes of demobilization are affected by the degree to which challengers provide themselves with enduring structures to maintain their solidarity (Tilly & Tarrow 2007: 97). Alliances do facilitate the development of shared identities, enabling groups with different views to establish shared meanings for collective actions that need to be undertaken (Kadivar 2013). Among accounts of the two specific outcomes of demobilization that we consider, democratization and counterrevolution, elite-driven approaches have mostly dominated in comparison to bottom up perspectives.[4] Concerning democratization, this has occurred within the transitology literature especially since the publication of O'Donnell & Schmit's hugely influential "Transitions from Authoritarian Rule" (Netterstrøm 2016: 397).

Most literature on democratization emphasizes institutional factors, related to the role of regimes and elites, in driving processes of democratization (Kapstein & Converse 2008). Among such factors, scholars have examined the compromises elites are able to find, the distribution of power between competing elites, the intervening role of the military in politics, and the influence of regional and international actors or brokerage among the elites (Higley & Burton 1989; Haggard & Kaufman 2016).

With reference to the events following the Arab Uprising, scholars have shown that inter-elite trust emerging in the aftermath of the 2010–2011 protests in Tunisia was a key promoter of the successful democratic transition (Hassan et al. 2020).

As was seen for processes of democratization, elite-driven approaches have prevailed in explaining counterrevolutions. Authors have explained counterrevolutions by linking them to international allies and pressures emphasizing the international tendency to attempt to overturn revolution (Bisley 2004). According to Allinson (2019), one strategy the counterrevolution in Egypt relied on was the integration of the post-Nasser Egyptian ruling elite with Gulf financial, and US security, networks. Others argue that, in the Egyptian case, antecedent military politicization gave military personnel more prominence than party apparatchiks in the political arena, making a military-led counterrevolution more likely to occur (Slater & Rush Smith 2016: 1478).

Bottom-up approaches to democratization have been also investigated. Studies have examined popular mobilization involving working-class uprisings (Rueschemeyer et al. 1992), alliances among workers, urban poor and peasants (Wood 2000), intellectuals (Kurzman 2008), ordinary people, and civil society (Bermeo 2003). Popular mobilization can undermine elites' power by lowering its legitimacy and support, as was the case in South Africa during the apartheid system (Schock 2005), by changing international alliances, or inducing a counter-elite favoring political negotiations (Wood 2000).

Scholarship on civil resistance has also shown the impact of nonviolent campaigns on democratization (Chenoweth & Stephan 2012; Celestino & Gleditsch 2013). Other research has suggested that unarmed collective actions is consequential for contentious democratization (Kadivar & Ketchley 2018).

Challengers' role in shaping counterrevolutions has been investigated less (see however Celestino & Gleditsch 2013, who argue that violent direct action makes transitions to new autocracies relatively more likely). We move along this direction and try to provide some insights on the role of challengers' alliances in shaping processes of democratization as well as counterrevolution. In the following paragraph, we illustrate how enduring alliances among challengers (or their lack) in the postrevolutionary periods may play a crucial role in the stabilization of challengers-elites interactions, conditioning the way they unfold, either to a democratic transition or into a counterrevolution.

Enduring alliances or fragmentation: a hypothesis

Identifying how challengers are able to keep up long-lasting alliances and mutual solidarity is a difficult task. Sustainable alliances are difficult to accomplish, especially for actors with diverse characteristics, interests, and identities. Furthermore, in non-democratic contexts, challenging groups are discouraged from building ties and relationships with each other due to the high risks associated with alliance-building processes.

Long-term alliances imply mutual trust, recognition, and solidarity, all features that are hard to find in former-authoritarian contexts undergoing postrevolutionary transition. At most, challengers may converge on specific and single issues, a strategy that does not grant long-term sustainability as it does not rely on shared feelings, final goals, and commitments (Melucci 1996).

Whether alliances may endure or not depends on various factors, including the actors' characteristics, and resources (see Edwards & McCarthy 2004). Organizations endowed with high levels of material and/or symbolic resources, such as status and popular legitimacy, may be less likely to engage in alliance building as they do not need to resort to alliances to gain additional resources (Diani 2015: 17–18 and 50–51).

Divergences in tactics, the attempt to strengthen one's own specific organizational profile, or that of securing a niche are all obstacles to enduring alliance building among groups.

Despite the difficulties in building long-term alliances, scholarly evidence shows how their structure is particularly important, in post-authoritarian contexts, for the consolidation of opposition forces driving political processes in emerging African democracies (LeBas 2011).

As argued by Kadivar (2018: 392–393):

> *unarmed popular campaigns that mobilize over a long period of time generate an organizational structure that provides a leadership cadre for the new regime, forges links between the government and society, and strengthens checks on the power of the post-transition government.*

The duration of mobilization matters for the durability of post-transition democracies, because popular campaigns typically require a solid organizational infrastructure to survive under repressive conditions (Andrews 2001).

In postrevolutionary settings, through alliance building, challengers may thus find a common terrain of shared objectives overcoming divergent interests, and possible reciprocal distrust. Alliance building among challengers are processes through which groups exchange different types of resources in pursuit of a common goal. This occurs as challengers realize they cannot afford to pursue their goals in total autonomy. The ultimate importance of alliances lies in their increased capacity to promote actions which are more likely to attract public attention, to be perceived as worthy, and to gain political legitimacy (Diani 2015: 25 and 50).

Fragmentation, in contrast, results from internal competition and polarization among challengers. Fragmentation may result from negative coalitions that can be defined as fractured elites lacking consensus over fundamental policy issues, and a weak commitment to democratic ends (Beissinger 2013). In the case of urban civic revolutionary coalitions, that is, regime change induced by rapid, leaderless, and socially diverse protest movements, the latter are prone to breakdown in the wake of regime change (Berman 2015). As argued by Berman (2015), this fragmentation becomes manifest in dysfunctional politics, as each fragment lays claim to the legacy of the revolution and the right to a dominating role in a new political order. Challengers become more and more autonomous from one another, drawing on different sources of support and a focus on diverging goals and tactics. Fragmentation, competition, and polarization are all processes likely to constrain the formation of broad collective identities, blurring any common understanding of collective actions.

These processes imply boundary reactivation among single groups who focus on their own specific interests and identities, a typical mechanism associated with phases of demobilization (Tilly & Tarrow 2015). Under such circumstances, fragmentation hinders processes of long-term coalition building necessary for a democratic transition, contributing instead to paving the way for a backlash to repression, leading to counterrevolution outcomes (Kadivar 2018: 395). Ultimately, when a fractured coalition of challengers fills up the postrevolutionary political vacuum, a counterrevolution may follow.

Continuing on from the aforementioned arguments, our hypothesis posits that the final outcomes of a process of demobilization is conditioned by the presence of long-lasting alliances among challengers. We contend that, in a postrevolutionary setting, the fragmentation of challengers' networks and prerevolutionary coalitions will tend to promote a counterrevolution. In contrast, when challengers are capable of establishing enduring alliances, we expect that the institutionalization and legitimization of challenger's alliances, and/or their support of the new elites, will contribute to political concessions and a transition to democracy.[5]

Data and methods

Case selection: controlled comparison

To empirically explore the aforementioned hypothesis, we focus on Egypt
and Tunisia before, during and after the 2010–2011 protests. In selecting
these countries, we followed the comparative method of paired controlled
comparison. This method implies the intentional selection of two or more
instances of a well-specified phenomenon that resemble each other in every
respect but one (Slater & Ziblatt 2013). Egypt and Tunisia were similar in
various aspects before the 2010–2011 protests, thus enabling us to assume
a certain number of rather constant background variables.

Both countries had Sunni Muslim majorities, a history of secular gov-
ernments and few oil resources. Moreover, both countries were marked
by costly food, unequal incomes, and high levels of youth unemployment
(Achcar 2013). Again, in both countries the 2010–2011 protesters drew
upon preexisting networks and fractures.[6] The 2008 protests in the textile
industry of Mahalla al-Kubra in Egypt had been characterized by wildcat
strikes not being supported by the official trade union, the Egyptian Federa-
tion of Trade Unions (ETUF), the only recognized trade union in Egypt dur-
ing Mubarak's regime. Social Islamic organizations and assistance networks
had also enabled individuals to become active participants in political life
in Egypt (Wickham 2002). Furthermore, in Egypt, organizations such as
Kifaya (*Enough!*), a network striving for reforms and change – including
organizations such as Journalists for Change, Doctors for Change, Youth
for Change, Workers for Change, Artists for Change – and the April 6th
Youth Movement (A6YM) were also active prior and during the Arab
Spring (Beinin & Vairel 2011).

In 2005, some members of the Muslim Brotherhood in Alexandria and
supporters of the Revolutionary Socialists also formed the National Alliance
for Change and Unions within universities. The same activists were later
amongst those who took part in anti-police riots that broke out after the
murder of the young activist Khaled Said in Alexandria in 2010 by a police
officer. As for Tunisia, the local groups of the major trade union, UGTT, had
been central during the local protests that occurred in 2008 in the Gafsa
mining basin (Beinin & Vairel 2011).

In Tunisia, next to UGTT, despite the political repression under both Ben
Ali and Bourguiba regimes, the ATFD also played an important role in the
opposition to the regime throughout the nineties and the new millennium.
Legalized in 1989, the ATFD focused its struggle against Ben Ali's state
feminism, against Islamism and rising conservatism (Debuysere 2018).

The ATFD supported several protests, some of which were sometimes
undertaken in cooperation with UGTT. These included those in Gafsa in
2008, despite women being largely underrepresented in the UGTT leader-
ship roles (Debuysere 2018). There were other parallels between the two

countries in the aftermath of the uprisings, until early 2013. They held free elections, assemblies were charged with drafting new constitutions, and Islamist parties won elections and assumed office (Hassan et al. 2020: 555). However, the two countries clearly showed divergent profiles after 2013. We will interpret such changes by examining one major difference observed in the two countries, that is, the type of networks and alliances built by challengers.

Paying particular attention to the timing and sequence of actions, we try to avoid the risk of endogeneity implied by a possible reverse causality given the high degree of interdependence of the actions and reactions involved. To do this, we examine the events at different time points in the empirical process ($t_0, t_1, t_2 \ldots t_n$). We therefore analyze the effects of protests observed at: t_1 which corresponds to December 2010 in Tunisia, and January 2011 in Egypt, on counterrevolution/democratic transition observed at t_2 corresponding to the military *coup* in Egypt in July 2013, and the passage of the new constitution in Tunisia in January 2014.

We explore the differences in the outcomes of demobilization observed in Egypt and Tunisia in light of the processes of long-term alliance building or fragmentation established by challengers in the two countries. While coalitions had already emerged between t_0 and t_1, we focus in particular on their developments during the period between t_1 and t_2.

Data sources

We rely on extensive fieldwork in both Egypt and Tunisia. Fieldwork in Egypt was conducted within the remits of a broad research project by the second author, analyzing Popular Committees and Independent Trade Unions in Egypt before and in the aftermath of the 2011 uprisings (Acconcia 2018). The specific data draws on fieldwork notes and on 58 semi-structured interviews carried out in Cairo and Mahalla al-Kubra with activists, workers and ordinary citizens involved in various groups.

The empirical evidence on Tunisia draws on various sources: first, data was collected within the framework of a broader research project, undertaken by the third author during fieldwork between February and May 2017. Data from the 2017 fieldwork consisted of 18 semi-structured interviews with selected participants across the popular neighborhoods of Ben Arous and Yasminette, in the suburbs of Greater Tunis, together with other eight in-depth interviews carried out in the neighborhood of Medinat Jedida with Tunisian youth. The Tunisian case study further draws on extracts from 100 interviews with women, undertaken between January and February 2011 in Kasserine (Zouhour and Nour) and Thala (Chennaoui & Baraket 2011).[7]

In addition to data deriving from this fieldwork, the second and fourth author conducted ten semi-structured interviews with Tunisian activists in Tunis and Sfax between April and July 2019.

As for Egypt, the interviews in Tunisia were all focused on the pre- and postrevolutionary period.

Egypt: the fragmentation of challengers' networks and the counterrevolution

The prerevolutionary coalitions in Egypt were largely based on common targets on single issues, for example, protests against police violence and anti-Mubarak sentiments (Clarke 2014). The lack of broader shared solidarities beyond these single issues marked the presence of a negative coalition (Beissinger 2013). Muslim Brotherhoods' supporters and secular activists, for example, participants within the *Kifaya!* movement, were already divided long before the 2011 uprisings in Egypt.

> *It is true that there was a participation within the Kifaya movement (2005) of some individuals within the Brotherhood and a common need to denounce Hosni Mubarak's repressive regime and his hereditary succession' attempts. However, we have always fought very different struggles,*[8]

a Kifaya activist who took part in Tahrir Square demonstrations explained. Those divisions were even clearer after the 2011 protests. The revolutionary coalition that preexisted the ousting of Mubarak – involving the Muslim Brotherhood, the army and many secular groups – had definitely split by the end of 2011.

Each group followed its own political track in the postrevolutionary period, consequently undermining Egypt's possible democratic transition. The Egyptian army tolerated the authoritarian regime of Mubarak as long as it was not detrimental to its corporate interests (Kandil 2012). In the circumstances of the 2011 uprisings, the military personnel briefly adopted the Muslim Brotherhood as a "civilian partner" (Abdelrahman 2015), later consolidating its powers. Among the major reasons for such a split was the "Muslim Brotherhoods' attempts to 'electoralize' contention and restrict a democratic transition to a process of negotiation, transaction and electioneering" (Ketchley 2017: 81–89).

The Muslim Brotherhood thought that no strong alliance with other partners was necessary to perform well in the November 2011 elections. According to a young activist who took part in Tahrir Square demonstrations the

> *Muslim Brotherhood left the youth of Tahrir alone in the streets in order to take part in elections and party politics paving the way for a marginal repertoire of street action.*[9]

Some attempts of alliance building had indeed been proposed through the establishment of the Democratic Alliance, founded in the summer of 2011, and composed of Freedom and Justice, the political party of the Muslim Brotherhood, and various other actors, such as some activists among the liberals and some old fashion leftist parties (e.g., *Tagammu*, *Karama*).

However, this was a short-lived experience and it vanished after the 2011–2012 parliamentary elections. The major basis on which such short-term alliance was built was a common fear of a possible backlash of authoritarian rules:

> At the 2012 presidential elections I decided to vote for Morsi not because he was my candidate but just because I did not want to see the Mubarak regime coming back,[10]

an activist who took part in Tahrir Square demonstrations said. Many opposition forces actually complained against the risk that the Muslim Brotherhood would control seat allocation in the 2011–12 parliamentary elections (Ketchley 2017: 88), as it eventually turned out.

Furthermore, the supporters of the Muslim Brotherhood mobilized throughout 2012 in order to oversee the electoral procedures for the Constitutional referendum (December 2012) and to protect the Presidential Palace of Heliopolis during the clashes following the presidential decree (November 2012).

Moreover, protesters became reliant on a 'focal day' (Friday prayers) to coordinate contention, and this had negative implications for coalition building (Ketchley & Barrie 2019).

Mobilization continued in 2013, on the occasion of the attacks on the Muslim Brotherhood headquarters (Moqattam, Manyal, and Mansour Street) between June 30 and July 9, 2013. Our interviewees mentioned that some members of the Muslim Brotherhood, recalling previous experiences of collaboration, were actually willing to hold discussions with other political parties between 2012 and 2013 in an attempt to build a political coalition with the Secular forces.[11]

> We tried to include other opposition forces within the Constituent Assembly. They were fragmented and stubborn. They tried to block the system, then they left the Assembly[12]

a supporter of the Muslim Brotherhood who took part in Tahrir Square and Rab'a Square protests explained.

All the secular political groups of the postrevolutionary period clearly showed a deep-rooted reluctance to trust the political ideology and leadership of the Muslim Brotherhood. Secular groups lacked trust in the army as well:

> The army is implementing the counter-revolution. However, our democracy should be better than the Western systems in which citizens are

> *forced to choose among two candidates who do not represent the grass-roots' demands,*[13]

a supporter of the Revolutionary Socialists highlighted. The political objective of secular groups had been more focused on trying to

> *bridge the gap between the tactics of new mobilizations and the more traditional methods of older social movement labour and rural organizing,"*[14]

a unionized worker in Mahalla al-Kubra stated. In particular, many of the secular oppositional groups aimed to join the textile workers' protesters of Mahalla al-Kubra and Cairo's protesters or workers' and agricultural workers' strikes with cyber-activism.

The army, on its side, broke solidarity ties with protesters right after the initial strategy of fraternization in early 2011 (Ketchley 2017). This change in strategy became clear in November 2011 in the Muhammad Mahmoud Street clashes and in May 2012 outside the Defense Ministry in Abbesiyya, when peaceful protesters were killed by soldiers. There were allegations of torture, forced virginity testing and repressive policing of street mobilization and labor agitation (Ketchley 2017: 75–77).

To restore order, the army was ready, overtly or furtively, to activate the police in order to defend the political elite. This renewed the alliance between the army and the police, or rather, between the Interior and the Defense Ministries, which was not as strong during Mubarak's time.

The alliance clearly surfaced during the rise of the 2013 anti-Muslim Brotherhood protests. This coincided with the split of the revolutionary coalition, as the secular forces of the *Tamarod* (Rebel), a petition campaign against the first ever elected president, a member of the Muslim Brotherhood, Mohammed Morsi, broke ranks from Islamist forces asking for Morsi's resignation.

With more than 461 worker protests between January and May 2013 (Alexander & Bassiouny 2014), the Egyptian secularist opposition pushed the military elite to be concerned about the consequences of the Muslim Brotherhood's neoliberal policies.

> *The Tamarod campaign unified the opposition's cliques: resentful citizens, public bureaucracy, the leaderless oppositions, families close to the National Democratic Party (NDP) and mercenaries,*[15]

an activist who took part in Tahrir Square demonstrations explained. These forces felt threatened by the winners of the 2012 presidential elections, the Muslim Brotherhood and their candidate Mohamed Morsi, and their depiction as violent terrorists was advanced by the army and national TV channels, arguing that they were not true and consistent democrats (Acconcia 2018).

The *Tamarod* movement further contributed to preparing the terrain for cementing the counterrevolution as it generated cross-class support for a process of political stabilization (Slater & Rush Smith 2016: 1476). This started with the authorities' 2013 ban on the Popular Committees, a strategic move to completely deactivate the mobilization of the Islamists.

According to our interviewees, the law banning the Popular Committees was first attempted by the authorities to identify the supporters of the Muslim Brotherhood, so as to later arrest or control them following the approval of the anti-terrorist law. The ban ended up putting all the activities, that since 2011 had been promoted by the Egyptian Islamist civil society, under strict control.

Interviewed activists who took part in Tahrir Square demonstrations stated that

> *The martial law gave to the police the right to kill. It allowed to a certain extent the society as a whole to be engaged in arbitrary actions.*[16]

And, as a supporter of the Muslim Brotherhood who took part in Rab'a Square demonstrations explained, the police can arbitrarily arrest, torture and violate human rights.[17]

On June 30, 2013, a huge demonstration took place in Tahrir Square. It was supported by the police, the judiciary, the Coptic Church, the public, and many private media. Former NDP politicians, the elders among Socialists and Liberals and intellectuals also participated. The demonstration was triggered by the *Tamarod* campaign.

As an activist who took part in Tahrir Square demonstrations said:

> *Many Egyptian followed Tamarod but we do not have to confuse Tamarod with the Coalition of Revolutionary Youth. In their public interventions Tamarod supported the army interferences. The revolutionaries would never have supported the police and a military coup, even if backed by the crowd.*[18]

Within 48 hours, President Mohammed Morsi was asked to resign. On July 3, he was arrested by the Presidential Guard and detained for months in a secret location without clear charges. The military coup took place on July 3, 2013, and engineered a restoration.

The backlash to repression entailed a protest law criminalizing opposition, the killing of 3,000 protesters in 2014, as well as the detention of those who instigated the mobilization against Mubarak (Ketchley 2017: 3–7). The Muslim Brotherhood, however, did not want a violent confrontation with the army.[19]

Instead, they organized peaceful demonstrations, marches, and flash-mobs throughout the country. In Cairo this happened in Rab'a al-Adawiya (Medinat Nasser) and al-Nahda (Giza); and peaceful protests were organized in Minya (Meidan Palace), Assyut (Meidan Omar Akram), Alexandria and Suez (Meidan Arbain).

Under such circumstances, the Brotherhood

> *turned increasingly to the activist base of the Islamist movement to act as 'police' – attacking demonstrators [from other factions], and in some areas taking over functions of maintaining law and order. These moves . . . inspired fears of both the collapse of public security and the emergence of a state of Islamist militias.*
>
> (Alexander & Bassiouny 2014: 282)

As for the allegations that the Muslim Brotherhood used violence toward the demonstrators, this has been constantly debated by the Secular and Leftist groups, some of which witnessed sporadic cases of violence.[20]

On August 14, 2013, repressive events escalated, followed by a sharp decline in popular participation (Grimm & Harders 2018: 7). The security forces attacked the Rab'a al-Adawiya and al-Nahda's encampments. Hundreds of people were killed or disappeared during these days.[21]

Those events clearly revealed the aggressive nature of the military-police takeover, paving the way for a less popular image of the Egyptian army, as confirmed by the very low turn-outs of the 2014 election, and again at the 2015 and 2018 ballots. Since the restoration of a military regime, no forms of dissent have been tolerated.

All forms of political activism, including youth and leftist groups, and political pluralism, were heavily suppressed. Public markets were closed, street vendors expelled and buildings in Cairo's downtown area were cleansed. The political leaders of the Muslim Brotherhood were stigmatized as full-fledged terrorists.

All the Brotherhoods' charities, including hospitals, NGOs, associations, and media outlets were either closed down or faced noticeable levels of repression, and the removal of former management. The Society of the Muslim Brotherhood, its political party the FJP, and the coalition defending the Morsi government's legitimacy were outlawed by the Egyptian courts. Islamist movements within the universities were heavily repressed (especially on the Al-Azhar and Ayn Shamps campuses in Spring 2014). As a Revolutionary Socialist activist who took part in Tahrir Square demonstrations argued, after the military coup, the alliance between the police and the army was instilling a sense of vendetta against the Islamists.[22]

Tunisia: alliance building and democratic transition

Compared to Egypt, Tunisian Islamist factions had a lesser influence within professional and workers' unions (Geisser & Gobe 2007; Ayari 2009; Chouikha & Geisser 2010). Nonetheless, in the years prior to the outbreak of the 2010 protests, some leaders of the main Islamist party Ennahda, outlawed and repressed during the 1990s, had already joined secular political parties and trade unions. This was especially the case within the largest

Tunisian trade union, the UGTT, which, unlike trade unions in Egypt, withheld a significant role in orchestrating protests and popular mobilization in the years preceding the 2011 uprisings (Pilati et al. 2019; Ketchley & Barrie 2019).

Likewise, the ATFD had cooperated with the UGTT during the 2008 workers' strikes in the Gafsa region, engaging also with Islamist opposition groups supporting the strikes (Debuysere 2018: 10). This had legitimized the ATFD, enabling the organization to engage as a key actor in the subsequent uprisings and political transition. These cross-group interactions, occurring in the prerevolutionary period, had contributed to the building of personal ties and levels of mutual trust amongst non-government affiliated and opposition actors, a crucial step for strengthening alliances during the postrevolutionary transition period (Hassan et al. 2020: 565). Arguably, prerevolutionary efforts by both Islamists and secularists in overcoming their mutual "distrust" were functional in paving the way for the later elite compromise in Tunisia (Stepan 2012: 92).

Despite the high levels of control and restriction applied by the Ben Ali regime, Tunisian civil society had been covertly active in creating spaces to advocate for socio-political and economic rights under the regime. The October 18 Coalition for Rights and Freedoms of the prerevolutionary Tunisia, for example, comprised of different secularist and Islamist opposition forces, was key in facilitating cooperation during the formation of the Constituent Assembly following the 2011 uprisings. The preexistence of covert associations was evident when, following the ousting of Ben Ali, the law was changed to make it possible to form independent civil society organizations. More than 4,000 of these organizations registered in just a few months, many of which had already been existing in secret (Feuer 2015).

At the first elections of the Tunisian Constituent Assembly in 2011 over 100 parties stood for election, representing the ample spectrum of ideologies of contestation to the old regime. These ranged from centrist, leftist, and environmentalist parties to the nationalist and pan-Arab fringes marginalized within the old regime ranks, to the Islamist factions outlawed during the Ben Ali rule. The elections were won by the Islamist party Ennahda that, in order to reach a parliamentary majority, had to form a coalition with two other parties: the social democrats of Ettakol and the centrists of CPR. This coalition was known as the "Troika."

The formation of the first postrevolutionary government could have led to a similar opposition fragmentation process as the one in Egypt. However, postrevolutionary events unfolded differently and fragmentation did not occur. The diverse coalition of Tunisian protesters that had deposed Ben Ali in January 2011 clashed over support and opposition to the Islamist-led Troika government (Berman 2015). Motivated by growing popular antagonism toward the new government's inability to recuperate a worsening economic recession, in 2013 a group of secular activists took inspiration from the growing *Tamarod* movement in Egypt to mobilize popular sentiment against the Islamist-led government. Initially mobilized over Facebook, the Tunisian *Tamarod* movement was mainly formed and led by disenfranchised and unemployed educated youth in Tunis's suburbs and other major inland cities.[23]

Adding to the Troika's inability to face price hikes, rising inflation and a downturn in investments, the government had to face mounting public anger following the assassinations of three political opposition leaders – Mohamed Lotfi Nagdh in October 2012, Chokri Belaid in February 2013, and Mohamed Brahmi in July 2013.

The assassinations were reportedly carried out by Salafi jihadists, as destabilizing actions against Ennahda's rule. As the Egyptian *Tamarod* movement radicalized, aiming to oust Muslim Brotherhood president Mohamed Morsi, the Tunisian version of the movement similarly advocated for the removal of the elected Islamist-led government as an appropriate solution to transit Tunisia out of the post-revolutionary political and economic *impasse*. Protesters against Ennahda's rule channeled the same disruptive repertoires of protest, sit-ins and demonstrations learned during the 2011 uprisings, operating outside the official routes of politics.

As stated by a young textile worker, UGTT member and local organizer in Medinat Jdida (Greater Tunis),

> we won't trust Islamists to take over. Tunisia is a progressive country. The revolution did not mean 'choose between a return to dictatorship and religious oppression.'[24]

When the Egyptian military staged a *coup* supported by a section of the Egyptian civil society to oust president Morsi, the Tunisian *Tamarod* movement initially supported the *coup*, encouraging a similar intervention to happen in Tunisia to curtail the Islamists' rise to power. However, the bloody repression perpetrated by the Egyptian army against pro-Morsi demonstrators soon made Tunisian opposition revert to its long-held position that the army should not intervene in politics, perceived as a threat to the State's secular assets (Holmes & Koehler 2020: 67).

After the August 2013 massacre in Rab'a Square, the violent crackdown that followed the military takeover of power in Egypt warned Tunisian secularists about the dangers of military intervention. The Tunisian *Tamarod* movement decided to adopt more institutional forms of political actions, opting for parliamentary involvement and representation through opposition parties, cherishing the culture of non-violence reinvigorated in the 2011 uprisings. This however, was not the only element that helped mitigate radical political contestation within the new arising political order.

As argued by Boubekeur (2016), the negative example of the Egyptian counterrevolutionary backlash was coupled with the emergence of "bargained competition" amongst secularists and Islamists. A political process aimed at avoiding further destabilization by steering toward a postrevolutionary settlement with old regime elites, and a national pacification process paved the way for the rise of a moderate conservative coalition in the *Nidaa Tounes* party (Call of Tunisia) led by Beji Caid Essebsi, former minister under Ben Ali's rule (Boubekeur 2016: 108).

The principle of "bargained competition" dominated the process of political compromise that began in August 2013, known as the Tunisian National Dialogue, focusing on overcoming the political impasse reached by the Troika government but also mindful of preventing the re-establishment of authoritarian mechanisms of State authority so vehemently challenged by the 2010 uprisings.

The National Dialogue process was led by the so-called Quartet of civil society organizations: the UGTT, the Tunisian Order of Lawyers, the Tunisian confederation of Industry, Trade, and Handicrafts (UTICA), and the Tunisian Human Rights League. Due to its prerevolution history of mild opposition, the UGTT had the moral leverage to pressure the competing elite camps, Ennahda in particular, to negotiate power (Hassan et al. 2020: 568).

The Quartet was not the only actor to provide a framework for coalition liaising among different sectors of the population. Another example of coalition building in the aftermath of the 2011 uprisings was the Coalition for Tunisian Women (CTW), a network of 22 feminist associations advocating for human rights, especially women's rights. This coalition bore upon prerevolutionary ties and included the TADW which, since its constitution in 1989, was active against both Ben Ali's State-promoted feminism, and against rising conservatism promoted by Islamist fields (Debuysere 2018).[25]

Female journalists and bloggers played a key role in the 2011 protests in Tunisia. In an interview with Amal Dhafouli, a 34-year-old activist in Sidi Bouzid, she argues that

> *I am independent but I collaborated a lot with the UGTT and other civil society organizations. . . . Women were during the revolution and until today one of the drivers of social movements in the regions.*

In the postrevolutionary period, between 2011 and 2014, CTW focused on the mobilization of women during electoral campaigns, advocating for women's equality in the new Tunisian Constitution, stressing the need for educational, health, welfare, social, and cultural rights for Tunisian women, especially in deprived and marginalized regions.[26]

> *Between 2011 and 2014, I have been working extensively with TADW, an association within CTW, focusing on education to democratic principles and against women discriminations,*

a CTW activist who took part in protests in Tunis argued.[27] CTW progressively decreased its activities among grassroots feminist associations, opting for increasing its institutional engagement in the Parliament and with the Presidential Council, proving crucial in bridging ties with political actors in the postrevolutionary period.[28] Within the Tunisian Islamist party Ennahda, several members had previously engaged in advocating for women's rights and had been involved in informal discussions with CTW on specific issues. As two CTW activists who took part in protests in Tunis stated:

> We met on several occasions with Ennahda members. However, they are still very conservative compared to our progressive views.[29] We discussed with the Islamists because the Tunisian political environment is less polarized than elsewhere.[30]

Due to growing levels of mistrust between Ennahda and CTW feminist activists, especially after the *Tamarod* campaign in 2013, and the murder of the Tunisian trade unionist Chokri Belaid in February 2013, the level of cooperation between CTW and the Tunisian Islamist activists constantly decreased.[31]

However, despite this and the difference in ideological stances, the fact that CTW and Ennahda activists did engage with each other at different times during the postrevolutionary phase of early democratic consolidating transition is evidence of the existence of trust building arenas and long-term alliance building between elites, activists, and all the diverse actors who were both part of the prerevolutionary coalition, resulting in compromise that is often necessary to setback authoritarian nostalgia. Interactions between CTW and Ennahda continued even after 2014, within the workings of the Individual Freedoms and Equality Committee, formed in 2017.

Their collaboration was evident in the reform of the Tunisian penal code,[32] especially during the debate on Law 58 on violence against women, approved in 2017.[33] As two CTW activists who took part in protests in Tunis stated:

> We worked with individual female members of Ennahda because they were receptive on the topic of violence against women.[34] Ennahda had a more moderate political agenda compared to the Muslim Brotherhood in Egypt and this helped for building up a dialogue with us.[35]

Alliance building between CTW and Ennahda was recently constrained by many divergent positions, as indicated by several CTW activists.

> On the reforms of the inheritance legislation [discussed in 2019], we have very different opinions compared to Ennahda female members.[36]

And other two CTW activists added:

> Nowadays, the collaboration between CTW and Islamist activists is very low and not visible.[37] One of the reasons for our mistrust is related to the continuous funding that Islamists receive from abroad.[38]

Mechanisms related to alliance building and fragmentation

Our analysis of processes of alliance building and fragmentation suggests that alliances surviving during the postrevolutionary period favor the democratic transition through a number of social mechanisms (Tilly & Tarrow 2015). First, they indicate the importance of trust building processes. During the transition period, high levels of reciprocal distrust within the Egyptian prerevolutionary coalition emerged. Opposition forces complained against the risk that the Muslim Brotherhood would control seat allocation in the 2011–12 parliamentary elections (Ketchley 2017: 88). Furthermore, as our interviews suggest, all the secular political groups of the postrevolutionary period clearly showed a deep-rooted reluctance to trust the political ideology and leadership of the Muslim Brotherhood.

Secular groups lacked trust in the army as well. Low levels of trust among members of the prerevolutionary coalition apparently contributed to the fragmentation of the coalition itself in the postrevolutionary period, a process that was instead mitigated in the Tunisian case.

Here, inter-elite trust in the postrevolutionary period was high (Hassan et al. 2020). While protesters did not fully trust Ennahda – particularly when the Tunisian version of the Tamarod movement advocated for the removal of the elected Islamist-led government – they engaged in processes of intergroup trust building. Thanks to the presence of some members of Ennahda who had engaged in advocating for women's rights and had been involved in informal discussions with the Coalition for Tunisian Women (CTW), reciprocal trust increased and fostered interactions among Ennahda and CTW. Interactions between CWT and Ennahda continued even after 2014.

A second mechanism that helps explain why long-lasting alliances favor democratic transition is the presence of a few brokers who became central in establishing links among challengers' organizations. Through overlapping memberships or ties with multiple groups, brokers can build cross-sectional alliances among groups themselves, despite their different interests, objectives, and ideologies. In other words, brokers produce new connections between previously unconnected individuals and groups (Tilly & Tarrow 2015: 31).

This had been the case of the Quartet in Tunisia, which played the role of "coalition broker" between secular and religious factions in Tunisia through the promotion of the National Dialogue process since August 2013. As mentioned, the UGTT had the moral leverage to pressure the competing elite camps, Ennahda in particular, to negotiate power (Hassan et al. 2020: 568). Likewise, the Coalition for Tunisian Women (CTW) was crucial in bridging ties with political actors in the postrevolutionary period, including Ennahda. Several individuals were also central brokers, liaising with different organizations, like Sana Ben Achour, who was, at the same time, a member of TAWD and of the Tunisian Human Rights League, one of the Quartet organizations.

A final mechanism that has possibly accounted for why long-lasting alliances favor democratic transition is ideological boundary deactivation. Long-term alliances are based on actors who engage in actions motivated by issues that go beyond pure interest in pursuing single themes and by shared collective identities. The prerevolutionary coalition in Egypt was largely based on targeting single issues, for example, protests against police violence and anti-Mubarak sentiments (Clarke 2014). Once that objective had been reached, members of the Egyptian prerevolutionary coalition followed their own paths and specific interests.

As mentioned, this involved the establishment of the Democratic Alliance, founded in Egypt during the summer of 2011, by the Freedom and Justice, the political party of the Muslim Brotherhood. Various other actors were also included, such as some activists among the liberals and some old fashion leftist parties (e.g., *Tagammu*, *Karama*). However, this was a short-lived experience.

Likewise, the army broke solidarity ties with protesters right after the initial strategy of fraternization in early 2011 (Ketchley 2017). Our results do not enable us to discuss the presence of a shared collective identity and ideological boundary deactivation among actors of the pre- and postrevolutionary coalition in the Tunisian case.

Several authors, however, argue that the ruling elite used a hegemonic discourse on "tunisianité" that may have supported alliances among challengers in Tunisia (Zemni 2016). "Tunisianité" may have contributed to creating consent for the ruling classes and to constituting a master frame capable of resonating with all different opposing views among challengers, thus deactivating their reciprocal boundaries. This is in line with those studies suggesting the role of ideological congruence as a crucial factor in fostering alliances (Croteau & Hicks 2003).

While we cannot assess the relative weight of the single specific mechanisms that our empirical evidence suggests, we believe that each of them can contribute to clarify why the construction of long-lasting alliances favors democratic transition. None of them implies the presence of long-lasting alliance building in itself, but all of them promote long lasting alliance building, especially if they occur together (for a similar argument, see Tilly 2004: 132). Trust and the sense of community can, in particular, co-exist as other studies focused on organizational dynamics in repressive contexts have shown (Bashri 2020: 8). In turn, the sense of community promotes boundary deactivation between specific groups and the construction of shared collective identities. Future studies may, however, investigate identity dynamics more broadly as our empirical evidence falls short in this regard. Furthermore, bridging actors may certainly contribute to increasing the levels of trust between groups that may have previously distrusted each other. This was particularly discussed with reference to the case of CTW and its interactions with Ennahda. As we have shown, CTW distrusted Ennahda initially. It was thanks to the personal ties that feminists of the CTW had built with some members of Ennahda, and their role as brokers, that CTW and Ennhada eventually built durable relationships that lasted at least until 2017.

Conclusions

This chapter aimed to address processes of demobilization in postrevolutionary contexts, attempting to clarify when the outcomes of demobilization are associated with processes of counterrevolution and when, in contrast, they lead to a concession of reforms and a democratic transition. Drawing on the Egyptian and the Tunisian events, we analyzed processes of alliance building and fragmentation in relation to the 2010–2011 Arab Spring protests and their aftermath, up to the military coup of July 2013 in Egypt, and the drafting of the new constitution in January 2014 in Tunisia. Our work aimed to complement studies focused on institutional arrangements and elite-driven approaches with a study on bottom-up processes leading to counterrevolutions or democratic transitions. In particular, we examined the role of long-term alliance formation among challengers.

Drawing on qualitative data from interviews held in both Tunisia and Egypt, the analysis confirms the significant role of long-term alliances in shaping different outcomes of demobilization. Whilst the literature on contentious politics often focuses on the role of alliance-building during the phase of mobilization, our contribution shows that the consolidation of challenger's coalitions is critical in shaping the process of demobilization.

This chapter discussed broad mechanisms of alliance-building. Furthermore, in line with those studies on coalition formation among SMOs (McCammon & Campbell 2002), we discussed the mechanisms of trust, brokerage and ideological boundary deactivation, highlighting why alliances may favor a democratic transition.

Notes

1 This chapter has been previously published in a different format in Political Studies Pilati, K., Acconcia, G., Suber, D. L., & Chennaoui, H. (2021). Protest demobilization in post-revolutionary settings: Trajectories to counter-revolution and to democratic transition. *Political Studies*. doi: 10.1177/00323217211034050

2 While for Tunisia, the term revolution is more commonly recognized among scholars, doubts are associated with the definition of events in Egypt (Gunning & Baron 2014). For the sake of clarity, however, we use the term pre and postrevolutionary period in both countries to refer to the years before and after 2010–2011.

3 For such reasons, counterrevolutions can also be referred to as involution or anti-democratic reactions.

4 For a complete account of theories of democratization, in addition to those centered on political variables discussed here, see Teorell, 2012.

5 See also Hassan et al. (2020). In contrast to these authors, we focus on networks rather than trust, as we believe that interactions and alliances are crucial factors for moderating the trajectories of demobilization.

6 For a broader discussion on the conditions that allowed those relations to flourish and the specific coalitions that have emerged in the prerevolutionary period in Egypt and in Tunisia see (Acconcia & Pilati 2021; Pilati et al. 2019).
7 The 100 interviews were part of a broader project in which Henda Chennaoui collaborated (Chennaoui & Baraket 2011).
8 Interview 7, Cairo.
9 Interview 25, Cairo.
10 Interview 42, Cairo.
11 Interview 53, Cairo.
12 Interview 55, Cairo.
13 Interview 34, Cairo. This interviewee clearly suggests the importance of regime dynamics. This emphasizes, as argued in the introduction, the need for integrating bottom-up approaches with elite-driven perspectives paying attention to regime dynamics, and a focus on the interactions between regimes and their opponents.
14 Interview 36, Mahalla al-Kubra.
15 Interview 5, Cairo.
16 Interview 7, Cairo.
17 Interviews 3 and 4, Cairo.
18 Interview 11, Cairo.
19 Interview 20, Cairo.
20 Interview 24, Cairo.
21 HRW, "All According to Plan. The Rab'a Massacre and Mass Killings of Protesters in Egypt," Human Rights Watch Report, August 2014. Available at: www.hrw.org/report/2014/08/12/all-according-plan/raba-massacre-and-mass-killings-protesters-egypt [Last accessed October 15, 2019].
22 Interview 7, Cairo.
23 Interviews 5, 6, 8, 13, 15, 16, 17, and 18, Yasminette and Medinat Jedida, Greater Tunis.
24 Interview 25, Medinat Jedida (Greater Tunis).
25 Interview 27, Tunis. During the interviews held in 2019, we focused on CTW, which includes the TADW established in 1989. These interviews show that the CTW had established various informal ties with members of the Quartet, especially within the Tunisian Human Rights League and the Tunisian Order of Lawyers. This occurred through a range of personalities, including Bochra Belhaj Hmida, lawyer and former president of TADW, Monia el Abed, law professor and member of TADW, and Sana Ben Achour, at the same time member of TAWD and of the Tunisian Human Rights League. As such, while we do not have empirical evidence from direct interviews with representatives of the Quartet, our interviews with CTW members partly compensate for this lack of data.
26 Interview 29, Tunis and Interviews 33 and 34, Sfax.
27 Interview 30, Tunis.
28 Interviews 30 and 31, Tunis.
29 Interview 27, Tunis.
30 Interview 31, Tunis.
31 Interview 27, Tunis.
32 Interview 28, Tunis.
33 Interview 31, Tunis.
34 Interview 32, Tunis.
35 Interview 31, Tunis.
36 Interview 27, Tunis.
37 Interview 29, Tunis.
38 Interview 33, Sfax.

References

Abdelrahman, M. (2015). *Egypt's long revolution protest movements and uprisings.* New York, NY: Routledge.

Acconcia, G. (2018). *The uprisings in Egypt: Popular committees and independent trade unions* (PhD thesis). London: Goldsmiths College, University of London.

Acconcia, G., & Pilati, K. (2021). Variety of groups and protests in repressive contexts: The 2011 Egyptian uprisings and their aftermath. *International Sociology, 36*(1), 91–110. doi: 10.1177/0268580920959246

Achcar, G. (2013). *The people want a radical exploration of the Arab uprising.* Berkeley, CA: University of California Press.

Alexander, A., & Bassiouny, M. (2014). *Bread and freedom, social justice. Workers and the revolution.* London: Zed Books.

Allinson, J. (2019). Counter-revolution as international phenomenon: The case of Egypt. *Review of International Studies, 45*(2), 320–344. doi: 10.1017/S0260210518000529

Andrews, K. T. (2001). Social movements and policy implementation: The Mississippi civil rights movement and the war on poverty, 1965 to 1971. *American Sociological Review, 66*(1), 71–95.

Ayari, M. B. (2009). Tolérance et transgressivité: le jeu à somme nulle des gauchistes et des islamistes tunisiens. *L'Année du Maghreb*, 183–203. doi: 10.4000/annee maghreb.569. Retrieved from https://journals.openedition.org/anneema ghreb/569

Bashri, M. (2020). Networked movements and the circle of trust: Civil society groups as agents of change in Sudan. *Information, Communication & Society, 24*, 470–489. doi: 10.1080/1369118X.2020.1859579

Beinin, J., & Vairel, F. (2011). *Social movements, mobilization, and contestation in the Middle East and North Africa.* Stanford, CA: Stanford University Press.

Beissinger, M. R. (2013). The semblance of democratic revolution: Coalitions in Ukraine's orange revolution. *The American Political Science Review, 107*(3), 574–592.

Berman, C. (2015). The durability of revolutionary protest coalitions? Bridging revolutionary mobilization and post-revolutionary politics. Evidence from Tunisia. Paper presented at Middle East Studies Association Meeting, Boulder, CO. Retrieved from http://chantal-berman.squarespace.com/s/negative-coalition-wp.pdf

Bermeo, N. (2003). *Ordinary citizens in extraordinary times: The citizenry and the breakdown of democracy.* Princeton, NJ, and Oxford: Princeton University Press.

Bisley, N. (2004). Counter-revolution, order and international politics. *Review of International Studies, 30*(1), 49–69. doi: 10.1017/S0260210504005820

Boubekeur, A. (2016). Islamists, secularists and old regime elites in Tunisia: Bargained competition. *Mediterranean Politics, 21*, 107–127.

Bratton, M., & van de Walle, N. (1997). *Democratic experiments in Africa regime transitions in a comparative perspective.* Cambridge: Cambridge University Press.

Celestino, M. R., & Gleditsch, K. S. (2013). Fresh carnations or all thorn, no rose? Nonviolent campaigns and transitions in autocracies. *Journal of Peace Research, 50*(3), 385–400.

Chennaoui, H., & Baraket, S. (2011). *Les Abandonnées de la Révolution – Étude des violences faites aux femmes à Thala et Kasserine lors de la répression de l'insurrection de décembre 2010-janvier 2011*. Uganda: Isis-WICCE Exchange Institute Alumni.

Chenoweth, E., & Stephan, M. J. (2012). *Why civil resistance works: The strategic logic of nonviolent conflict*. New York, NY: Columbia University Press.

Chouikha, L., & Geisser, V. (2010). Retour sur la révolte du bassin minier. Les cinq leçons politiques d'un conflit social inédit. *L'Année du Maghreb*, 415–426. doi: 10.4000/anneemaghreb.923. Retrieved from https://journals.openedition.org/anneemaghreb/923

Clarke, K. (2014). Unexpected brokers of mobilization: Contingency and networks in the 2011 Egyptian uprising. *Comparative Politics, 46*, 379–397.

Croteau, D., & Hicks, L. (2003). Coalition framing and the challenge of a consonant frame pyramid: The case of a collaborative response to homelessness. *Social Problems, 50*, 251–272.

Debuysere, L. (2018). Between feminism and unionism: The struggle for socioeconomic dignity of working-class women in pre- and post-uprising Tunisia. *Review of African Political Economy, 45*, 25–43. doi: 10.1080/03056244.2017.1391770

Della Porta, D. (1995). *State and political violence, social movements, political violence, and the state: A comparative analysis of Italy and Germany*. New York, NY: Cambridge University Press.

Diani, M. (2015). *The cement of civil society. Studying networks in localities*. New York, NY: Cambridge University Press.

Durac, V. (2019). Opposition coalitions in the Middle East: Origins, demise, and afterlife? *Mediterranean Politics, 24*, 534–544. doi: 10.1080/13629395.2019.1639969

Edwards, B., & McCarthy, J. (2004). Resources and social movement mobilization. In D. A. Snow, S. A. Soule, & H. Kriesi (Eds.), *The Blackwell companion to social movements*. Oxford: Blackwell.

Feuer, S. (2015). *Beyond Islamists & autocrats: Post-Jasmine Tunisia*. Washington, DC: Washington Institute for Near East Policy.

Geddes, B. (2011). What causes democratization. In R. E. Goodin (Ed.), *The Oxford handbook of political science*. Oxford: Oxford University Press. Retrieved from www.oxfordhandbooks.com/view/10.1093/oxfordhb/9780199604456.001.0001/oxfordhb-9780199604456-e-029

Geisser, V., & Gobe, E. (2007). Des fissures dans la "Maison Tunisie"? Le régime de Ben Ali face aux mobilisations protestataires. *L'Année du Maghreb*, 353–414. doi: 10.4000/anneemaghreb.140. Retrieved from https://journals.openedition.org/anneemaghreb/58

Grimm, J., & Harders, C. (2018). Unpacking the effects of repression: The evolution of Islamist repertoires of contention in Egypt after the fall of President Morsi. *Social Movement Studies, 17*(1), 1–18. doi: 10.1080/14742837.2017.1344547

Gunning, J., & Baron, I. Z. (2014). *Why occupy a square? People, protests and movements in the Egyptian revolution*. London and New York, NY: Oxford University Press.

Haggard, S., & Kaufman, R. R. (2016). *Dictators and democrats: Masses, elites, and regime change*. Princeton, NJ: Princeton University Press.

Hassan, M., Lorch, J., & Ranko, A. (2020). Explaining divergent transformation paths in Tunisia and Egypt: The role of inter-elite trust. *Mediterranean Politics, 25*(5), 553–578. doi: 10.1080/13629395.2019.1614819

Higley, J., & Burton, M. G. (1989). The elite variable in democratic transitions and breakdowns. *American Sociological Review, 54*(1), 17–32.

Holmes, A. A., & Koehler, K. (2020). Myths of military defection in Egypt and Tunisia. *Mediterranean Politics, 25*(1), 45–70. doi: 10.1080/13629395.2018.1499216

Kadivar, M. A. (2013). Alliances and perception profiles in the Iranian reform movement, 1997 to 2005. *American Sociological Review, 78*(6), 1063–1086.

Kadivar, M. A. (2018). Mass mobilization and the durability of new democracies. *American Sociological Review, 83*(2), 390–417. doi: 10.1177/0003122418759546

Kadivar, M. A., & Ketchley, N. (2018). Sticks, stones, and molotov cocktails: Collective violence and democratization. *Socius: Sociological Research for a Dynamic World.* doi: 10.1177/2378023118773614

Kandil, H. (2012). *Soldiers, spies, and statesmen, Egypt's road to revolt.* London: Verso.

Kapstein, E. B., & Converse, N. (2008). *The fate of young democracies.* Cambridge: Cambridge University Press.

Ketchley, N. (2017). *Egypt in a time of revolution: Contentious politics and the Arab spring.* New York, NY: Cambridge University Press.

Ketchley, N., & Barrie, C. (2019). Fridays of revolution: Focal days and mass protest in Egypt and Tunisia. *Political Research Quarterly, 73*, 308–324. doi: 10.1177/1065912919893463

Koopmans, R. (2004). Protest in time and space: The evolution of waves of contention. In D. A. Snow, S. A. Soule, & H. Kriesi (Eds.), *The Blackwell companion to social movements* (pp. 19–46). Oxford: Blackwell.

Kurzman, C. (2008). *Democracy denied, 1905–1915: Intellectuals and the fate of democracy.* Cambridge, MA: Harvard University Press.

LeBas, A. (2011). *From protest to parties: Party-building and democratization in Africa.* Oxford and New York, NY: Oxford University Press.

McCammon, H. J., & Campbell, K. (2002). Allies on the road to victory: Coalition formation Between the suffragists an and the Woman's Christian Temperance Union. *Mobilization: An International Quarterly, 7*(3), 231–251.

Melucci, A. (1996). *Challenging codes. Collective action in the information age.* Cambridge: Cambridge University Press.

Netterstrøm, K. L. (2016). The Tunisian general labor union and the advent of democracy. *The Middle East Journal, 70*(3), 383–398.

Pilati, K., Acconcia, G., Suber, L., & Chennaoui, H. (2019). Between organization and spontaneity of protests: The 2010–2011 Tunisian and Egyptian uprisings. *Social Movement Studies, 18*(4), 463–481. doi: 10.1080/14742837.2019.1567322

Rueschemeyer, D., Stephens, E. H., & Stephen, J. D. (1992). *Capitalist development and democracy.* Chicago, IL: University of Chicago Press.

Schock, K. (2005). *Unarmed insurrections: People power movements in non democracies.* Minneapolis: University of Minnesota Press.

Slater, D., & Rush Smith, N. (2016). The power of counterrevolution: Elitist origins of political order in postcolonial Asia and Africa. *American Journal of Sociology, 121*(5), 1472–1516. doi: 10.1086/684199

Slater, D., & Ziblatt, D. (2013). The enduring indispensability of the controlled comparison. *Comparative Political Studies, 46*(10), 1301–1327. doi: 10.1177/0010414012472469

Stepan, A. (2012). Tunisia's transition and the twin tolerations. *Journal of Democracy, 23*(2), 89–103.

Tarrow, S. (1989). *Democracy and disorder: Protest and politics in Italy, 1965–1975.* Oxford and New York, NY: Clarendon Press and Oxford University Press.

Tilly, C. (2004). *Social movements, 1768–2004*. Boulder, CO: Paradigm.

Tilly, C., & Tarrow, S. (2007). *Contentious politics* (2nd ed.). New York, NY: Oxford University Press.

Wickham, C. R. (2002). *Mobilizing Islam: Religion, activism, and political change in Egypt*. New York, NY: Columbia University Press.

Wood, E. J. (2000). *Forging democracy from below: Insurgent transitions in South Africa and El Salvador*. Cambridge: Cambridge University Press.

Zemni, S. (2016). From revolution to Tunisianité: Who is the Tunisian people? Creating hegemony through compromise. *Middle East Law and Governance, 8*(2–3), 131–150.

4   The gendered effects of the war

Poverty and displacement of Syrian women in Lebanon

In this chapter we want to shed light on the condition of women and girls in displacement displaced within Lebanese borders during the ten years of war in Syria. Their lack of protection from both a legal and material point of view, with the consequence that many aspects of their lives are worsened by displacement are the main focus of the interviews.

The principal outcome of the study is the fact that like violence against women in the study led by Heise (1998), the violation of the reproductive rights and the right to parenthood of Syrian refugees in Lebanon are not only connectable to individual wrongs or a cultural vacuum, but perpetrated on many different levels, between which women find spaces and measures to put forth their personal will, despite their extremely negative construction as "helpless" and "undeveloped" women who live in the MENA region.

Introduction

As we enter the tenth year of the Syrian civil war, initiated as a peaceful protest against the authoritarian regime of Bashar al-Assad, whose family has been ruling the country since 1971, approximately half of the prewar Syrian population has been displaced internally or externally. Almost six million Syrians have found refuge in neighboring countries. Among the states hosting large numbers of Syrian nationals, Lebanon is the one in which living conditions are at their most precarious, forcing refugees into a limbo of extreme poverty and uncertainty that permeates all aspects of their lives.

Although literature has been increasingly involved with the gendered effects of displacement, the majority of studies have seemingly been conducted from a legalistic point of view or aimed at providing facts and figures about the refugee population taken as a whole. Another trend in the outputs of scholars, international organizations, and NGOs is to create a unified "women and children" category, inside which it remains unlikely to encounter a satisfying answer concerning the conditions in which women really live or decide to have children, for example, and which factors influence them or condition them to do or not do so.

DOI: 10.4324/9781003293354-5

The methodological approach taken in this study is a mixed methodology based on an analysis of the existing literature on the topic, including existing qualitative and quantitative research, with reference to the existing international human rights standards that are provided by the instruments concerning asylum seeking, refugee rights, and reproductive rights. In particular, by adopting Heise's ecological model of analysis (1998), the focus of the analysis is on the consequences of the Lebanese environment on the reproductive choices of Syrian women. A series of in-depth interviews was also conducted to support the qualitative analysis and to give these women back their voices and agency. On the one hand, it would be extremely blinkered and unjust to say that women are always capable of transforming their wishes and desires into practice in reproductive matters, as in any other matter, since many women suffer from a variety of unspeakable violence that saddle them with a never-ending series of burdens accentuated by their poverty conditions.

Reproductive rights are particularly at stake in light of the extremely unfavorable conditions in Lebanon. Extended xenophobia, the privatization of the health system, and the absence of a legal framework of protection, along with a personal lack of awareness about the available services, as well as family dynamics, religious background, and traditions, are all potential sources of violation of women's reproductive rights. Society at large, from the micro level to the macro level, reinforces the oppressive environment in which Syrian women make decisions regarding their bodies.

According to the principal international human rights instruments and the provisions contained therein, women worldwide are entitled to access information about sexual and reproductive rights and health, as well as access to services and medicines. They are entitled to give their consent after being granted that their choices are made on the basis of reliable information and that when they choose to be mothers, they are to be protected in the name of the right to a safe and healthy pregnancy.

In the words of Ross and Solinger (2017), women are entitled to all the means that can grant them the possibility to exercise their "right not to have a child; right to choose to have a child; and their right to parent children in safe and healthy environments." Syrian women who are forcibly displaced in Lebanon are not granted any of these rights. Instead, they are pushed and pulled in different directions: by their partners, their employers, their extended families, and by the Lebanese national authorities that refuse to afford them protection, from any point of view. What renders their situation extremely vulnerable is that it originates from the fact that the host state cannot be defined as such in legal terms, since Lebanon refuses to ratify the Refugee Convention of 1951 and its Protocol for specific political reasons.

Equally, the Syrian situation is rendered even more complicated by the strong prejudices that are spread among the Lebanese population. The general Lebanese xenophobic discourse, as seen in many of the studies collected for the purpose of this research, is that, if Syrian refugees cannot afford to sustain the upbringing of that many children, they should not have as many as they do. Health providers harshly criticize Syrian women for the number of children they have, but they do not realize that families in Syria were more numerous because of the public nature of healthcare, social welfare, and education, which allowed parents to have more children without worrying about providing for their basic needs. As a result, in Lebanon today, Syrian women are struggling to find free or affordable reproductive healthcare and are mistreated for being pregnant. A lower percentage of women were found to be using family planning methods (34.5%) compared to before the conflict in Syria (58.3%). In addition, the Lebanese health sector has prohibitive costs when compared with the incomes of displaced Syrian families, the vast majority of whom live in conditions of extreme poverty, according to the Vulnerability Assessment for Syrian Refugees in Lebanon (VASyR), compiled by the UNHCR. Most refugees also live outside Beirut, which means that they can access health clinics only by taking transportation, which they cannot always afford. The same goes for contraceptives, which they either cannot afford in the first place, or they cannot afford to implant, even when given for free. This contributes to the high fecundity of the Syrian population.

The Lebanese setting also has the fault of not permitting abortion unless it is to preserve the health of the mother if it is placed at risk by the pregnancy itself. Combined with the poor socioeconomic conditions of Syrian refugees, abortions are not accessible for refugee women, since the price of an illegal abortion varies between 150 USD and 2,400 USD. Therefore, Lebanon, as a host country, fails to be a country of reproductive justice.

From a micro-level point of observation, personal histories and relationships with partners are also sources of inducements. Some inducements derive from the patriarchal nature of Syrian society even before displacement, as the vast majority of the women interviewed cited their dependence on their husbands' desires when it comes to having or not having a certain number of children or taking contraceptive measures.

There is also a fear of undergoing divorce or polygamous marriages. The same thinking applies to the growing and disturbing practice of marrying girls at a very young age, a phenomenon accentuated by the hardships of displacement, especially because parents are unable to provide for their children and are in fear for their own lives. Young and unmarried girls are extremely vulnerable to sexual violence, harassment, and abuses, perpetrated by both co-nationals or Lebanese, which puts their reproductive health and reproductive choices in an extremely precarious situation. Many unmarried women are forced to engage in transitional sex in order to survive and make ends meet. Displacement has had notable consequences for the reproductive patterns of younger married couples and younger women in general.

A study by ABBAD has shown that the Syrian population has not been immune to changes in traditionally constructed gender roles, although they have not always come at no cost. For many women, breadwinning and providing for the family comes at a very high price, including increased domestic violence, intimate partner violence, or forced relegation to the domestic sphere as a means to restore the status quo of pre-conflict society. Nonetheless, displacement has also had the positive effect of making certain partners more reasonable when presented with the economic and legal hardships of the Lebanese context as a motive to have a reduced number of children.

Significantly, Syrian women who have not changed their minds, and who still intend to have a high number of children, face a series of difficulties from the beginning of the pregnancy through to postnatal care. Syrian women have extremely low rates of routine check-ups and are forced to find coping mechanisms in response. In the past, when the Syrian health system was not unavailable as it is now, women displaced near the border used to go back to Syria, especially if their children had postnatal issues, where they could be treated free of charge. Today, this scenario is hardly ever available, and returning to Syria is very dangerous for women. As a result, many are forced to rely completely on informal healthcare facilities, mostly put in place by NGOs, while others address their problems with the help of Syrian informal health workers, who are unauthorized to exercise their profession in Lebanon. In sum, from within the Syrian community displaced in Lebanon, the general impression is of an extremely high number of people who have been left with nothing and who are now deprived of the one thing that is fundamental at a general level: family.

The ecological framework of analysis

Syrian refugee women make their reproductive choices on the basis of interpersonal, institutional, infrastructural, and experiential constraints and inducements. This chapter aims to explain how the Lebanese framework is abusive of its own women and queer citizens, which dramatically limits the choices of migrant and refugee women.

From the abusive institutional level derives a series of infrastructural and economic inducements, while Syrian women are themselves at the intersection point of their religions, their families, and their political positions.

The model proposed by Lori Heise (1998) can be applied to this research, and from it, we can understand that the negation of Syrian refugee women's reproductive rights and their right to parent has different levels of causality. We theorize the existence of four different levels at which violence against women is constructed and perpetuated: *personal history*, which is the innermost dimension that influences women; the *micro-level*, meaning that women are influenced by their kinships, partners, and closest family; the *meso-level*, which encompasses the influence of the women's work environments, for example, or the extended families of the peer groups in which they participate; the *exo-level*, which refers to structural violence and how institutions and social structures perpetrate violence; and the *macro-level*, which comprises the general attitudes that permeate the culture at large.

Interestingly, according to Yasmine and Moughalian (2016), relegating these constraints solely to the interpersonal sphere means relegating reproductive injustice to a cultural scheme, whereas in reality, it is embedded in a wider picture that principally involves the attitudes of all the actors that interact with Syrian refugees throughout their presence in Lebanon – from NGOs to the government, from local communities to health providers.

The macro-level: religious and social constraints

At the macro-level, the experience in Lebanon for many displaced Syrian women and men has been accompanied by widespread racism, sexism, and xenophobia. As explained in the first chapter, relations between the two countries have a complicated political background, which is concretized in a pervasive hostile attitude toward the refugee population. The same cage of constraints must also include all the stigmas that prevent women from reporting violence, rape, and sexual assaults, which have a high impact on their reproductive health and choices.

This level should also include the religious beliefs of these individuals. Since the vast majority of displaced women are Muslims, it is worth dedicating some space to analyzing the relationship between religion and motherhood in the geographical area we are investigating, mostly because it determines some of the wider social patterns to which we refer. As already stated, religious beliefs and backgrounds have a tremendous impact on the reproductive choices of women worldwide, particularly because religious institutions are embedded in societies and have a powerful influence on the minds of the people. Almost 90% of the Syrian refugee population in Lebanon are practicing Muslims, mostly Sunni; hence, there is a need to analyze how the religious dimension influences the reproductive rights of these women (Minority Rights Group International 2019).

Islam is an extremely diverse religion, and its interpretation of the principles contained in traditional religious sources differs across the Muslim world. Abortion, as in other religious contexts, is highly debated. According to Hessini (2007), there are four main schools of thought on the matter:

- *Abortion is allowed.*
- *Abortion is allowed under certain circumstances.*
- *Abortion is disapproved.*
- *Abortion is forbidden.*

Support for abortion and the belief that life begins at ensoulment are based primarily on the following Qur'anic verse (23, 1–50), which discusses the different stages (semen, blood clot, bones, and flesh) of fetal development:

Man We did create from a quintessence [of clay]; then We placed him as (a drop of) sperm in a place of rest, firmly fixed; then We made the sperm into a clot of congealed blood; then of that clot We made a [fetus] lump; then We made out of that lump bones and clothed the bones with flesh; then We developed out of it another creature. So blessed be Allah the Best to create!

Abortion was practiced until the beginning of the 20th century, when the western medicalization of birth started. During the 1960s, despite the initial backlash toward the decisions of many states of the MENA region to implement family planning policies, recalling the Qur'anic provision that Allah will provide for all, two pan-Islamic conferences decided to approve the use of birth control in order to avoid pregnancies for those families that lacked the financial resources to provide for more children than those already born (Hessini 2007).

Today, the position on abortion continues to change. Many states allow the practice of abortion if it is to protect the life of the mother, for economic reasons, or to protect the offspring of the couple that may be endangered by the new pregnancy. However, the terms under which abortion can be practiced are often debated, as they refer to the necessity of ending the pregnancy before ensoulment, which, according to different schools of thought, takes place 40, 90, or 120 days after conception (Hessini 2007).

The crucial point here is whether faith and freedom can coexist concerning reproductive justice. To what extent does the faith of the individual and of the health providers influence women? In particular, faith-based organizations, which are present in most displacement contexts, fail to provide women with a full range of contraceptive methods; indeed, some refuse to provide them to unmarried individuals because of the stigma attached to pre-wedding sexual intercourse. Abortion is also one of the most absent services when it comes to family planning.

On a more personal level, the majority of the interviews realized for this study dramatically show the moral and shaming implications that women experience when they consider abortion. Some do not even conceive of contraception or abortion as an option. Here, the Islamic narrative perpetuates a strong societal and religious construction around the role of the mother; as in many other religions, the family is portrayed as its founding block. Islam presents an entire discourse celebrating the role of women primarily as mothers. One example of this attitude is the parallel of *motherhood* with *Jihad*, meaning that motherhood is the means to follow Allah's way, like men have to follow *Jihad*.

This discourse is contained in a *hadith* in which a woman interacts with a prophet who explains to her why women are not required to participate in *Jihad* and why they are to be considered martyrs if they die during pregnancy or breastfeeding, since raising a good child is the most virtuous deed a woman can do, and it is subject to compensation even in the afterlife, stressing that *"Heaven lies beneath the feet of mothers"* (Hidayatullah 2014).

Based on this construction of motherhood, women who even think about going through the abortion process are victims of societal shame and self-criminalization. In turn, they feel the need to give adequate moral justification to those performing the procedures and to themselves (Fathallah 2019).

Being a Muslim woman means having a strong engagement in the procreative function, but it is fundamental not to do so before marriage. Premarital sex is strictly forbidden for both sexes, but especially for women, because it would lead to the children being claimed by someone who is not their righteous father, and a man taking care of someone who is not his offspring is considered unethical (Badissy 2016). Nevertheless, physicians performing abortions seem to be more prone to practice abortion on unmarried girls than on those women who pursue abortion against the will of their partners, mainly because they want to save the reputation of the young girls (Fathallah 2019). According to UNFPA, unmarried women in developing countries face the pressure of dealing with a "socially unsanctioned" pregnancy, which increases their vulnerability (Yasmine & Moughalian 2016).

Cultural factors also have a strong influence. As Abouelnaga (2018) reports, women are conceived as mothers in social narratives to such an extent that it is considered incomprehensible not to become a mother at some point in a woman's life. In the words of the author:

> We women who live in Arab societies have been raised with the notion that our lives and bodies do not belong to us. And despite the disparities of our lived realities, philosophically and materially, we are held to the same societal expectations when it comes to motherhood. It is as if my cultural and social contribution is limited to me taking pride in the photos of my children that I must show to whomever comes my way. Otherwise, I become suspected of not offering anything of value to society.

In this sense, Syria has been a country with high birth rates and has preserved such rates by being a country through a healthcare system that was affordable to all. Before the war, the country was a middle-income state where about 3% of national expenditure was invested in providing competitive health services for all. Although there was the option of consulting private professionals according to one's economic possibilities, maternal healthcare was universal, as confirmed by the growing rate of births in clinics attended by skilled personnel (Bashour & Abdulsalam 2005).

Challenging the idea of protection

Today, it is increasingly difficult to be a Syrian refugee in Lebanon. Economic, social, and legal pressures make it difficult to define the country as a safe "refugeehood." The refugee situation continues to be precarious, with extensive humanitarian and developmental needs. On a general level, the 2018 VASyR reports that in recent years, the households of Syrian refugees have shifted from extended family households toward a more "nuclear model," with an average of five members per family. The gender ratio appears to be balanced between males and females, of whom over half (54%) are below 18 years of age. A small percentage of households are administered by women (around 19%), and almost 60% of families have at least one family member who has some kind of particular need, defined as "having physical or mental disability, chronic illness, temporary illness or injury, a serious medical condition, and/or needing support in basic daily activities" (UNHCR et al. 2019: 21).

The protection space for Syrian refugees in Lebanon is strongly affected by the policies enforced in 2015. Strong restrictions affect access to the country, rendering it almost impossible to cross the frontier, except for a few exceptional circumstances approved by the Ministry of Social Affairs (MoSa). Furthermore, it is extremely difficult to obtain legal status, which limits the lives of adults and those of their children, who cannot be registered at birth. As the 2018 VASyR reports, 73% of those aged 15 years or older have no legal status within the country, and households living in nonpermanent structures have a higher ratio of undocumented members.

Until September 2017, parents needed to have a residence permit in order to register their children through a complicated process. The requested documentation was as follows: a notification of birth from the hospital or midwife and a birth certificate from the Mukhtar. The birth had to be registered with the competent local civil registry office (Noufous) within the first year of life. The procedure is complicated not only because of the lack of legal status of the parents, but also due to the difficulties in accessing the necessary documents (UNHCR et al. 2019: 27).

The consequences of the absence of legal status can be divided into two macro-categories: limitation of freedom of movement and lack of access to justice. The first category refers mostly to the fact that people who are not in possession of legal status fear for their security and try to avoid crossing at checkpoints. The Norwegian Refugee Council (NRC) reports that the majority of male refugees are afraid to pass military checkpoints because they fear arrest and harassment. Especially in the north, the vast majority of those interviewed by the NRC limit their movement to Tripoli, the major city and location of the most important UNHCR and Lebanese offices, because of the checkpoints at Halba and El Bireh. Many also report that they have experienced incidents at checkpoints throughout the country; in some cases, their documentation has been damaged or confiscated without a valid reason (Norwegian Refugee Council 2014).

The military forces randomly select buses and cars and check whether there are undocumented male Syrians who are travelling. According to NRC, most of the time, those who have proof of their legal status are left free to continue their journey while the others are arrested, and although arresting women and children is not a usual practice, it is common for both women and men to be mistreated. Sometimes, the fear of crossing checkpoints leads to the choice to refuse to seek health services; they are thus confined to the closest facilities, which are usually pharmacies or private doctors who charge for their services, instead of reaching hospitals or clinics where they could be economically helped by humanitarian actors.

The second consequence of the lack of legal status is the impossibility of accessing justice. Many do not report abuses by the police for fear of being arrested. This leaves them exposed to exploitation, abuse, and violence. The NRC reports one case of a father whose daughter was kidnapped; on reporting it to the police, the man was arrested, and the search procedure was never initiated (Norwegian Refugee Council 2014).

Among the people interviewed for this study is the family of an eight-year-old boy from El Bireh who was hit by a car driven by a Lebanese man and left paralyzed and unable to speak because of extensive brain damage. The family received no redress for the accident, and the driver was never charged for not assisting the boy in the first place, even though he was later identified by the Syrian family.

Exposure to abuses also extends to renting. When seeking shelter, undocumented refugees are dependent on their landlords, who can evict them without notice because they cannot sign a tenancy agreement. Landlords can also increase their rent (for apartments as well as for the place where they put their tents) by cutting off their electricity supplies. Some face abuse by their employers, who may refuse to pay for a job already done without fear of punishment because the undocumented refugee cannot contact the police (Norwegian Refugee Council 2014).

Shelter for refugees

Government policy prevents Syrian refugees from establishing formal camps in Lebanese territory. Consequently, refugees live in cities, villages, or informal camps. The last two years have seen a shift from mostly residential to nonresidential forms of sheltering. Shelters are mostly chosen on the basis of the financial possibilities of the household, but also based on their proximity to the employer (if there is one) or to family. The majority of tenancy agreements are verbal; only a small minority of the population reports possessing signed contracts. The average cost can vary from 58 USD for a non-permanent structure to 220 USD for a residential lodging in an urban area. The VASyR reports that the majority of the population lives below the poverty line and in substandard shelters, with leaking roofs, rotting walls, no windows or doors, non-functioning water or sewage systems, and inadequate electricity installations. Part of the population also lives in overcrowded structures, with less than 4.5 square meters of space per person. These shelters can sometimes be dangerous, not only due to their specific structure, but also due to the environment in which they are placed (UNHCR et al. 2019).

In the camp where most of the interviews were conducted for this study, several tents have caught fire because of the poor condition of the electrical system. We also heard reports of a girl, aged around three, who died a few meters away from her tent because she fell into a sewage hole that was uncovered, and of a six-year-old who died after a car hit her in front of the door of the garage in which her family lived, which faces a heavily trafficked road. Many of the refugee population are forced to change their housing structures at various times. These changes are made in light of the household's economic possibilities (i.e., those who lose their source of income are forced to move from apartments to cheaper accommodations, such as garages or tents, and vice versa) or due to problems with their landlords. Moreover, a new trend has been evident over the last year: militarily enforced evictions, in which individuals or families are removed from their homes or the lands they occupy, against their will (Operazione Colomba 2019).

Before the military evictions, Human Rights Watch reported mass evictions of Syrian refugees by the local administration within a paralegal framework, thereby non-respecting the principle according to which a court authorization has to be issued by a judge before conducting an eviction. The NGO cited these forced evictions as outcomes of nation-based or religious-based discrimination, depending on the specific cases (Human Rights Watch 2018a: 38). These new military mass evictions, however, cannot be placed in the same category. The decision by the Higher Defense Council, a military body, was welcomed by the Lebanese foreign ministry, which said that the demolition drive would prevent refugees from permanently settling in Lebanon.

In July 2019, Al Jazeera reported that a series of demolitions had affected some non-formal camps in Arsal. According to the author of the article, the military had previously given the refugees a warning at the beginning of June, intimating them to modify the structure of their housing structures so that no concrete was present within the buildings.

At the end of June, almost half of the hard shelters in Arsal were destroyed by their own inhabitants (Khodr 2019); this was followed by a military raid that tore down any structures that were not compliant with the regulation of "concrete free tents," leaving many without any decent housing alternatives (Chehayeb 2019).

These evictions, whether conducted unlawfully by the municipalities, by the landlords, or by the military, have had a massive impact on the refugees. Many have had to leave their property behind, along with the already-paid rent for that month, while at the same time having to pay for a new deposit, which is not always possible due to hardship in accessing work for the mostly undocumented Syrians.

They are also forced to relocate their children from their schools or to spend more in order to continue sending them to the same one. Surveys of forced evictees show that many of them have responded to the situation by borrowing money (Human Rights Watch 2018b).

The impact of the Lebanese healthcare system on refugees

Lebanon has a very complicated healthcare system that balances the presence of public and private actors. The system can be described as pluralistic due to the mixed public – private actors in both service provision and funding to access these services. Almost half of the Lebanese population is covered financially by the National Social Security Fund, governmental schemes, or private insurers. The remaining half does not benefit from any formal insurance and is covered by contracts of the Ministry of Public Health, which encompasses low-cost consultations for those affected by chronic diseases and vaccines. The private sector dominates healthcare service delivery channels; 80% of hospitals are private. Some facilities that belong to NGOs are available (World Health Organization 2017). A wide array of health services is provided by NGOs, religious charities, and political parties, which have access to extensive resources, especially for middle- and low-income citizens.

The prolonged Syrian influx has placed a significant burden on the Lebanese healthcare system (World Health Organization 2017). According to the 2018 VASyR, most of the Syrian population has received medical assistance at least once during their presence in Lebanon. Most of these were forced to visit private doctors. The main barriers to accessing primary healthcare or hospitalization in Lebanon are linked to economic factors. Some of the interviewees in the VASyR reported that they could not afford the treatment or the cost of transportation to the facility, or that they could not leave a deposit and were therefore denied acceptance. As for funding, only a small percentage of the population reported the presence of health insurance that fully covered the necessary procedures. The remaining assert that they receive cash assistance (mostly from NGOs) or fee waivers. NextCare, for example, is the UNHCR program, which waives 75% of medical expenses in partnering hospitals and clinics (UNHCR et al. 2019). On a general level, a study conducted by El Arnaout et al. (2019: 3) reveals the poor health conditions of the Syrian population:

> Studies have indicated that non-communicable diseases (NCDs), mental health disorders including post-traumatic stress disorder (PTSD) and depression, as well as communicable diseases (CDs) such as Cutaneous Leishmaniasis, have been noted to be the most prevalent cases observed in refugee settings.

The impact of decreasing aid

For the last eight years, international and national aid has proven to be fundamental in enhancing the coping mechanisms of Syrian families. However, in the last year, aid supporting the Syrian crisis response has decreased dramatically, to the point that only 33% of the required funding has been transferred to the Lebanese Crisis Response Plan (LCRP). The decreasing incoming aid has concretized the impossibility for many Syrians to make ends meet, since for many, this was their only source of income at some point in their presence in Lebanon. In particular, fiscal cuts to the World Food Programme's Multi-Purpose Cash Assistance Program (MCAP) and to the UNHCR program have left many Syrians exposed to extreme poverty (Mhaissen & Hodges 2019).

According to the VASyR report, on average, 68% of households had at least one working member during the last year, reporting an average of 13–14 working days per month, with significant variances from region to region; for example, the northern part, the Akkar Valley, and the southern part only averaged around 10 working days per month. Female-headed households are the poorest, with a significant percentage of "zero working members" within the family. Syrian refugees are legally permitted to work in sectors in which they were traditionally engaged before the outbreak of the conflict – agriculture and construction – but recompense is not always fair (UNHCR et al. 2019).

An enlightening interview conducted by the Brush and Bow journalist's collective gives us a glimpse of the problematic and unequal labor market into which some Syrians are forced to enter. The interviewee was a woman, Randa. She speaks about the complicated and exploitative economy into which Syrian families are thrown because of the absence of legal protection. She reports that her personal experience, as with many others, has been linked to agreements between human smugglers and the Shawish (a person elected by the other refugees of the camp, who is responsible for communicating with the international organization for aid/food and with the landlord for rent-related problems and to intervene in disputes within the camp itself) of informal camps, who pays for newly smuggled individuals to be brought into their refugee camps in order to have a larger workforce. A cycle of exploitation is imposed by the everlasting debt owed to the Shawish, for building a tent for the family, for providing the materials to build it themselves, or for "rescuing them" from the smugglers. Randa describes a system of forced labor as a means to repay thousands of USD to the Shawish. Along with the labor exploitation, Randa denounces a whole series of mistreatments: the confiscation of documents for the UNHCR registration process, the consequent theft of the cash sum given to every refugee to start their settlement in Lebanon, or the confiscation of the goods provided by humanitarian actors and their resale for a higher price, among many other forms of abuse (Brush & Bow 2019).

In this scenario, the picture of Syrian livelihoods is overwhelmingly negative: they are heavily dependent on humanitarian aid as well as earnings from seasonal employment in the agricultural and farming sectors. Wages are extremely low (less than 2.90 USD per person per day) and irregular in arriving, while working hours are long. A large proportion of young children also work in the agricultural sector (Al Zoubi 2019).

To face these difficulties, Syrians enforce various coping mechanisms, especially when it comes to procuring food and providing earnings in the medium – long term. Almost the entire population lives in a situation of food insecurity, meaning that food provision is the priority of many families. Half of households reduce meal portions or the number of meals throughout the day. It is also common for adults to adopt food restrictions (i.e., reduced food intake) in order to properly feed their children. Expensive food is avoided, giving priority to cereals, potatoes, and vegetables.

As for livelihood coping mechanisms, the most used measures are as follows: saving as much as possible and spending the lump sum during periods with no income, selling household goods and furniture, reducing non-food expenditures, removing children from school, marrying the children off before the age of 18, or involving them in the labor market. Some of these are only emergency measures (UNHCR et al. 2019).

Reality and myth in the "return to Syria"

It has been claimed that in the last three years, around 173 Syrian refugees have returned home from the three neighboring host countries with the most displaced people (Turkey, Lebanon, and Jordan). Such returns seem to be the result of a growingly hostile environment, as well as the movement of warfare in Syria. At a societal level, there has been a rising call for returns; now that the war in Syria appears to be over, local populations are full of discontent toward the refugee population, which has concretized a deleterious public rhetoric, as seen in increased xenophobic tendencies and discrimination-driven incidents (Içduygu & Nimer 2019).

The same patterns can be found within Lebanese society. While all the political actors are united in calling for international support to mitigate the crisis, Foreign Minister Gebran Bassil was already calling for the deportation of Syrians in 2013: "Lebanon should stop receiving refugees unless for exceptional cases and the Syrians already in Lebanon should be deported." This anti-refugee rhetoric has continued, becoming extremely harsh after the fights against Jabat Al-Nusra militants in the refugee camp of Arsal, when the local Shi'a authorities, and Hezbollah in particular, repatriated almost 10,000 people over the Syrian border (Içduygu & Nimer 2019: 7). During the last year, the grip of the Lebanese government has become even tighter, pushing the Syrians toward a so-called "safe and dignified return." The idea of "safe return" has its roots in the good neighborhood principle; at the same time, it can be read through the lens of the Lebanese rhetoric on the refugee status of Maya Janmyr: if there is a safe place to return to, the Lebanese authorities are excused in not legalizing the presence of the refugees and in putting all their efforts into relieving the country from the burden of the displaced (Janmyr 2018).

Unfortunately, the reality of voluntary returns is far from the "safe and dignified return" perpetuated by the authorities. As Habbal (2019) states, Syrians sometimes have to choose between a life of misery, or a return to the unknown. Eventually, a variety of push factors end with some families deciding to go back to their homeland, whether they are legal, economic, or social in nature.

The legal status granted to Syrian refugees is extremely precarious, with the possession of valid residency being an exception rather than the norm. The high fees and the reticence of the Lebanese authorities to confer legal status to the displaced have made it almost impossible for Syrians to provide for their families or to live without the fear of arrest or mistreatment. In 2017, Human Rights Watch reported the death of five Syrian refugees in Lebanese military custody (Human Rights Watch 2018b). This followed an NRC report on the difficulties caused by the lack of legal recognition, including abuses on behalf of the police, systematic arrests at checkpoints, incapacity to find a job, and fear of reaching health-providing structures such as hospitals and clinics, among many other more minor effects (Norwegian Refugee Council 2014).

The legal framework has a direct consequence on the livelihoods of Syrian refugees, who are forced to implement coping mechanisms to fight food insecurity, bad sheltering conditions, and the impossibility of access to a semi-private health system. Furthermore, the New Arab reports that, during the last 12 months, Lebanon's General Security Directorate has begun to enforce very restrictive laws on Syrian employment. Throughout the country, small stores run by Syrians have been systematically closed, and the documents of those working there have been confiscated and never returned. One testimony reports that:

> *Everyone, every store, that had an employee or worker, per person, [had to pay] one million liras [about 650 USD] or had 48 hours to pack their things and to stop work.*
>
> (Kranz 2018)

The pressure to return is also placed on refugees through the policy of dismantling Syrian housing structures when they are found to be noncompliant with a longstanding law on the imperative absence of concrete in tents, which had never been enforced previously (Operazione Colomba 2019). In April 2019, the Lebanese government announced a systematic campaign of dismantlement in the town of Arsal to avoid permanent refugee housing communities. The refugees themselves were forced to tear down their houses and take their possessions; otherwise, the army would have intervened, and no object could have been saved from within the structure. Aid groups in the area estimated that the measure impacted around 3,000 shelters and some 15,000 refugees (Habbal 2019).

A further push factor is linked to the social environment in which Syrians live, which is mainly concretized in some form of harassment. In

January 2019, the Daily Star reported that Syrian refugees in Arsal had been attacked by Lebanese protesters complaining about the economic environment. Many complained about not seeing any improvement in their living conditions but have still not gone back because it is not safe (Salloum & Hodges 2019: 17). As an outcome of these various push factors, many have chosen to return to Syria. The Lebanese government has made an agreement with the Syrian authorities to send refugees back to "secure areas." As early as 2018, Major General Abbas Ibrahim, the head of the General Security Agency, stated that "there are contacts with the Syrian authorities about thousands of Syrians who want to return to Syria. The stay of Syrians in Lebanon will not go on for a long time. There is intensive work being done by the political authorities" (Reuters 2019).

The conception behind the call for returns is that the Government of Syria, guided by Assad, now controls the vast majority of the territory, and in many of them there is no active fighting. Contrary to this presumption of the coincidence of absence of warfare and safety, international actors have denounced the absence of basic services in these "pacified areas," as well as the structural violence and the lack of respect for basic human rights. To make an informed decision and return voluntarily, refugees need to be granted a full array of information about the safety of the area to which they are returning, the status of their housing, and the risk of forced conscription or detention at the border after their return. Most of this information is neither currently available nor reliable (Salloum & Hodges 2019).

The existing procedure involves consulting the Government of Syria (GoS) to obtain information about the legal conditions of those willing to return. The UNHCR says that a necessary premise is to "engage and seek partnership with GoS and host governments on return to Syria, outlining conditions required for sustainable returns in safety and dignity, while furthering advocacy against coerced returns" (UNHCR 2018: 10). The actors involved in the voluntary returns are multiple, and most of them work informally. The most important is the GSO Directorate, which collects the names of those who want to return and sends them to the Syrian authorities to accept them. They are the coordinators of the operations. It takes about two months for the application to undergo the procedure, and it takes only three days from the acceptance notification to the actual departure. Hezbollah has also opened nine centers for voluntary returns in the east and south of Lebanon. There is no insight into their internal processes or the authorities with which they coordinate to guarantee a safe return. At an informal level, local actors have also engaged in facilitating returns to Syria. Returning through tribal connections is considered safer than the governmental option, since the tribal leaders accompanying the returnees also follow them into Syrian territory and avoid arrests at the border or forced conscription during the journey (Salloum & Hodges 2019).

Returns organized through the GSO cost about 200 USD per person, while those organized by local committees cost around half this sum. During the period that precedes the return, refugees are particularly vulnerable to economic exploitation, as they try to settle their debts and usually sell their housing materials and furniture, which may result in them accepting too little in order to settle as much as possible (Salloum & Hodges 2019). Human rights defenders have raised concerns over whether these returns can be considered viable options and whether they are truly voluntary. At a conference at the Camera dei Deputati, the Italian Parliament, Operation Colomba asserted that the returns were proceeding not because of Syrian guarantees of internal safety for its citizens, but because of the ever-worsening situation in Lebanon. Furthermore, there are other preoccupying factors and testimonies of people who have returned and lost their homes because of Law No. 10, a legislation passed by the Syrian government in 2018, giving property owners only 30 days to claim the ownership of their property. Returning men between 18 and 49 are also at risk of being forcibly conscripted, having to endure arrest, or paying a fine for not serving in the military previously (Habbal 2019). These are also the concerns of Syrians considering returning to their homeland. Physical protection is lacking; the Syrian Network for Human Rights and Human Rights Watch have reported extrajudicial killings and forced disappearances, as well as movement restriction, extortion, and gender-based violence, such as rape and sexual assault (Salloum & Hodges 2019).

The worsening social and economic conditions for refugees in Lebanon are not the only reason behind the returns to Syria. There have been episodes of forced returns. Habbal (2019) reported on an episode in the city of Zgharta, where the GSO confiscated the documentation of those applying for residence permit renewal and forced them to sign the forms for voluntary returns to Syria. The same misconduct has also occurred in other areas where refugees were misled into signing for returns while applying for registration, only to be informed that they were going back to Syria within a month.

Lebanon: a country for women?

It does not take long to understand that Lebanon is not really a "champion" state when it comes to defending women's rights and ensuring the application of the international legal frameworks of gender equality and reproductive rights. A clear example of its attitude toward gender issues is its citizenship law.

In Lebanese society, women's citizenship is not as valuable as that of men. Lebanese women are granted political participation, but they are not entitled to pass this citizenship on to their own children, who acquire their father's instead. The country relies on *jus sanguinis*, but without considering the blood of women to be equally valuable to that of men. Consequently, the children of Lebanese women and non-Lebanese men are forced to face institutionalized discrimination in their legal residency, employability, property ownership or inheritance, education opportunities, and access to health services (Human Rights Watch 2018b).

Paradoxically, children born out of wedlock, which is the only exception in which the transmission of nationality is maternal and not paternal, are afforded a better social environment than those who have a Syrian or Palestinian father, which correlates fully with the aforementioned discourse regarding asylum and how the Lebanese constitution refers to non-nationals.

Lebanon ratified the Convention on the Elimination of All Forms of Discrimination Against Women (CEDAW) in 1996. However, amendments by the Lebanese government to Article 9, Paragraph 2, and Article 16, Paragraphs 1 (c, d, f, g) and 3 have since refuted the purpose and objectives of CEDAW. The rejected articles related to the personal status laws and nationality rights of female citizens (Avis 2017). Through these reservations, the Lebanese state effectively denied women the same rights as men in instances of marriage, divorce, and family matters; it also upheld the ban on Lebanese women from passing their nationality to their husbands and children, which is functional in perpetuating the maintenance of personal status laws under the control of the religious courts rather than their civil equivalents (Salameh 2013).

Furthermore, as voiced by many women rallying on the streets during the uprisings that began in October 2019, Lebanese society keeps women on the margins of every decision-making process that affects their lives, beset by overlapping and intersecting discriminations that derive from the overarching political, legal, social, and economic structures (Geha 2019).

The patriarchal structure of society has handed control of the entire Lebanese public sector to men. The absence of women in any decision-making process is to be attributed to the fact that sectarianism has divided society into a state governed by customary laws rather than codified ones, rendering even parliamentary and public representation male-dominated. This is because the system of power sharing in the state is based on the various sects and religious communities, which are typically dominated by male members of leading families. This leads women, and men, to be governed from birth to death as a function of the sect into which they are born or married (Avis 2017; Geha 2019).

In turn, Lebanese law fails to protect or enhance the basic human rights of women, but also of refugees, economic migrants, queer people, persons with disabilities, working-class people, transgender people, those living with HIV, and sex workers. Lawmakers are almost always cis-men of influence, status, and power, and the laws, which do not reflect the human rights discourse, are the outcome of their positions alone. Indeed, the abuses take place in the gap between the international human rights discourse and national laws, which are the result of the liberty given to each nation state to implement (or not) the provisions contained in the major conventions and multilateral treaties (Rola & Batoul 2018).

Consequently, assessing or at least describing the condition of reproductive justice advancement in Lebanon requires an in-depth analysis of different layers of the social structures in which women are embedded. In the following sections, we will analyze how these layers of abuse are constructed, how they impact the lives of Syrian women, and which other elements contribute to determining the reproductive choices of Syrian women displaced in Lebanon. The spectrum of sources that negatively impact or limit women's access to reproductive justice is wide. In the Lebanese context, women face stark situations when it comes to bodily self-determination and the right to dignified parenthood.

The struggle for reproductive justice has as its principal barriers: women's legal status, marital status, refugee-migrant status, access to healthcare, restrictions on women's physical mobility, unavailable or expensive transportation, social stigma, sectarian stigma, and interactions with family members, spouses, and religious figures (Rola & Batoul 2018).

The consequences of poverty (not only for women)

From a reproductive justice perspective, poor women worldwide are not considered fit to become mothers. As we have already explained, the narratives of nation states often portray white women with economic resources as "elected mothers." The reproductive justice movement claims that the capitalist nature of our societies penalizes women, who, rather than being denied the opportunity to parent, should be granted economic assistance (Ross & Solinger 2017). From this premise, it is logical that the economic status quo of a family is a great determinant in the reproductive choices of women or of kinship. In displacement scenarios, female-led households are extremely vulnerable, especially when they are expected to support themselves and the wider family. Employment opportunities are scarce in the context of displacement and may even be reduced further when traditional gender biases come into play.

Beyond gender-based violence, traditional cultural practices may also limit women's access to basic goods and services as well as livelihood opportunities in a displaced context. An assistance program distributing staple food items in a refugee camp, for example, may inadvertently follow male-oriented leadership structures that subsequently limit its ability to reach women and girls. The same reasons may underpin preferential access to healthcare and instruction for boys and men. Studies (Bilgili et al. 2017) have shown that young women and girls usually find jobs such as housekeepers, cooks, and hairdressers not far from their residence site, or they are in charge of small businesses built informally within their homes. This renders their income low and insecure, which sometimes leads women to adopt sex as a transactional means of gaining support for themselves or their families. Although one branch of feminists argues that this practice should be called "empowerment prostitution," according to Pittaway and

Bartolomei, it is not only improper but also culturally biased, since in many displacement contexts, such women must endure shaming and discrimination, persevering purely for their own survival. The term "survival sex" more properly represents the nature of the activity; the empowering aspect does not apply to most displacement environments, and it hardly applies to western societies (Pittaway & Bartolomei 2018).

Naturally, the use of sex as an "exchange good" strongly impacts the reproductive pattern of women, and we cannot speak of reproductive justice or sex work as a free choice, which means that women who engage in survival sex are denied their reproductive rights. These women do not choose their partners, and they engage in intercourse that – especially in displacement – is not always accompanied by contraceptive methods or protection against sexually transmitted infections. Contraceptive/protection methods are also strongly dependent on the income of the families or the women themselves. They may be unavailable near the housing structure in which the women live, so obtaining them (either through purchasing them or from a specific organization that provides them for free) could be conditioned by the possibility to afford transportation. Moreover, even if contraceptive and protection methods are available in proximity to refugees, they may not be free or could represent a disproportionate expense in the context of the family budget.

Another dramatic consequence of poverty, although it is fair to say that personal beliefs are also part of the process, is the increasing ratio of child marriages, which, especially in displacement, are mostly driven by poverty and preoccupation with the general safety of girls. While child marriage itself is a harmful and dangerous practice, displacement adds further elements of vulnerability since, most of the time, the marriage lacks legal recognition for both the wife and the children that are born as a result. Further challenges are represented by the "stateless" status that some children face having been born out of illegitimate marriages, or the possibility of abandonment of the underage brides or their forceful involvement in polygamous marriages, leaving them unprotected (Karasapan & Shah 2021).

The UN's Sustainable Development Goal 5 calls for "achieving gender equality and empowering all women and girls." Among the many operational necessities, one of the most important sub-targets of the goal is to eradicate forced marriage worldwide, which, in 82% of the cases, involves underage girls (Karasapan & Shah 2021). The reasons behind the increased early marriages in displacement contexts derive mostly from the uncertainty in which most of these families live. Many marry their daughters because they fear for their physical safety.

Since girls are vulnerable subjects who can be victims of sexual violence, marriage can be considered a reparation for what has been done to them previously through displacement; it is a manner of maintaining the family's honor. Young girls can be the targets of harassment and gender-based violence; therefore, families believe that marriage can be a form of granting girls protection (Freedman 2015). This elucidates the ways in which heteropatriarchy delineates marriage as protective from the physical and sexual harassment of strangers. In Lebanon, the same reasons lead many women to (re)marry into abusive relationships or polygamous marriages. Economic factors also play a fundamental role in this context. Marrying a girl at a young age can be a way for families to relinquish the necessity of providing for them, since it is common practice for the husband and his family to provide for the bride.

There have also been reports of families who used their daughters as a good to be exchanged in order to gain directly from the marriage itself (i.e., through the dowry). Furthermore, parents who are not certain about their own survival may want to marry their daughters off to be certain that if they do not survive, the girls will at least have a family and a safety net (Freedman 2015). The consequences on the reproductive choice rights of girls and adolescent women are evident.

Data from 15 developing countries reveal that adolescent women under the age of 17 are far less likely to receive skilled prenatal and delivery care than women between the ages of 19 and 23. Moreover, the children of young mothers are likelier to be born prematurely and at low birth weights; they are also more likely to be stillborn or die within the first four weeks of birth (Green & Merrick 2006).

In the refugee camp where most of the research had been conducted, not all of the girls were married adolescents. One notable case was that of a two-year struggle by a Syrian family to protect their two underage daughters from the advances of a 50-year-old Lebanese man. The man first tried to get engaged with the older daughter, who was about 16 at the time, and the parents were forced to let her marry a young Syrian boy to save her from the man. Her younger sister, who was 14 at the time of the interviews, was almost kidnapped in front of her school, allegedly by people close to the Lebanese man, who had decided to get her since he could no longer go for the older sister, who was now a married woman. Two other girls in the same refugee camp were both married by the age of 14.

They reported that it was their mother's wish to see them both marry in the difficult circumstances of displacement. The impact on their reproductive lives is self-evident; they have both abandoned schooling and started families instead. As of June 2019, the older daughter was 17 and a half, with a young son who was around a year and a half old and another baby on the way, while her sister was almost 16, with a son who was around one and a half. Poverty also determines whether women can access healthcare, and the quality of the healthcare they are provided. As an outcome, prenatal cures and the ratio of maternal morbidity change disproportionately between those who have economic resources and those who have none. Global, regional, and country-level estimates of maternal mortality show a clear connection between high rates of maternal mortality and poverty (Green & Merrick 2006).

Violence as a systemic issue

The elimination of violence is fundamental to achieving reproductive justice. Violence is exercised at a medical state level on those who do not follow traditional binary gender assignments and on those who do not respect socially constructed and accepted behaviors. On an individual and interpersonal level, domestic violence, intimate partner violence, street harassment, rape, and a variety of legally sanctioned behaviors are also significant. In Lebanon, domestic violence is treated differently according to the sectarian origins of the women involved. The laws on personal status implicitly sanction domestic violence, but they also depend on the sectarian origins of each woman. Crucially, many women are not in a position to divorce violent husbands because of the lack of monetary responsibilities for these men, as well as the high probability of not winning custody of their children. Intimate partner violence is not redressed, as it is widely accepted that a wife's duty is to render herself sexually available to the husband, who can ask for a divorce if the woman is not fulfilling her marital duty. Article 503 of the Lebanese penal code reinforces the idea that rape cannot be committed when perpetrated under wedlock.

The article defines rape as "forced sexual intercourse against someone who is not his wife by violence and threat." This article of the penal code also enforces the idea of penetration as a unique act of sexual violation, reinforcing the social stigma around men who are raped but do not report these cases to the patriarchal authorities for fear of being emasculated and associated with the socially constructed feminine sex (Rola & Batoul 2018).

Violence against women is a continuum, but due to the displacement framework, it can assume some particular features and can be rooted in different grievances. For women and children whose male counterparts have participated in war, there is evidence to confirm that domestic violence increases after conflict when combatants return home. In the Serbian context, for instance, research has demonstrated that the hyperviolent masculinities constructed during wartime, in combination with the psychological wounds left by the fighting, turn to domestic violence (Bradley 2018). In strongly patriarchal societies, if before displacement women have experienced a change – mostly induced by the conflict – in their interaction with their partners or with their family context in general, they are expected to return to their traditional roles, in order to avoid the risk of an escalation of violence to reaffirm the antecedent dominant structure (Bradley 2018).

As Atkinson et al. (2005) highlight, "for women, breadwinning can be dangerous," since married men who have few resources to offer, or fewer resources than their wives, are likelier than their resource-rich counterparts to use violence. In this way, violence serves as compensation for their shortage of resources. In cases of domestic violence, women may be helpless depending on different factors. Domestic violence may be a socially accepted practice, meaning it is normalized or not perceived as a crime; there can be no interest in addressing a phenomenon that is considered ordinary.

One study conducted on the postwar Bosnian environment showed how societal beliefs influence the manner in which domestic violence is addressed. The evidence suggests that in societies that believe domestic violence to be a private affair, violence is underreported and normalized even by police officers, who do not intervene in such contexts (Muftić & Cruze 2014).

Another element that may impede women from seeking help or redress in violent situations is the fear of having to sustain the economic burden of the family in displacement alone if they leave their husbands. Freedman (2017) also explains that the lack of documentation forces women to endure domestic violence because they fear that any report could trigger the repatriation of the family member who is perpetrating violence against them, or they fear for their own freedom.

Sexual violence is a plague that is widespread among displaced women, but it is not a peculiar characteristic of displacement; it is an everyday threat to women's safety worldwide. That said, women are particularly vulnerable in camps where shelters are not secure and easily accessible, and because of the necessity to walk extensive distances to find basic items:

> *Imagine living in a refugee camp where you are too scared to go the toilet or being subjected to sexual harassment on a daily basis in your host community because of your gender or identity. This is the terrifying reality for hundreds of thousands of women and girls and LGBTI refugees around the world.*
> (Catherine Murphy, Amnesty International 2016)

The lack of male protection, as in households led by women, increases the risk for sexual violence, although even its presence does not guarantee safety, as sexual violence may come from intimate partners. One study from 2014 highlighted how pervasive the phenomenon is: around 21% of women from 14 different states reported being victims of sexual violence while being forcibly displaced (Robbers et al. 2016).

Women are not only victims of sexual violence during times of conflict; they are also victims of these abuses while on the move. Smugglers usually take advantage of their position and violate women who travel unaccompanied or even in the presence of their husbands, fathers, or brothers, although the presence of men is a powerful deterrent in some contexts (Robbers et al. 2016). There is also the disturbing phenomenon of sexual violence perpetrated by aid workers, police, and officials.

According to a study conducted in 2012, almost 70% of female migrants and refugees who entered Europe reported experiencing sexual violence, often perpetrated by professionals, which indicates that – compared to 11% of reports of sexual violence among European citizens – these crimes were committed by taking advantage of the extreme vulnerability of these women, especially refugee women (Robbers et al. 2016). Sexual violence has devastating consequences, as it can result in unintended pregnancies or in the contraction of diseases, such as HIV and sexually transmitted infections (Robbers et al. 2016).

A lot of women in refugee camps from Jordan contracted urinary infections because of abstinence from using toilets during the night (Amnesty 2013). In sub-Saharan Africa, sexual abuses are so widespread that women who engage in migration, whether voluntary or forced, usually take contraceptive pills as part of their survival kits (Amnesty 2016). In addition to physical consequences, sexual violence has a dramatically negative impact on women's mental health, and it can lead to post-traumatic stress disorder as well as anxiety and depression (Robbers et al. 2016).

Moreover, in many cultures, sexual violence remains a taboo subject and is often underreported. In many cultures, stigma, and shame associated with rape can lead to underreporting of cases, social rejection, suicide, or murder of women and girls by family or community members. Humanitarian personnel may also not be fully equipped to address the situation (Hynes & Cardozo 2000).

In Lebanon, rape and other forms of sexual violence disproportionately target refugee/migrant women and people with nonnormative gender expressions. While the phenomenon affects both citizens and noncitizens, it is much harder, if not impossible, for noncitizens to find recourse to justice, report incidents, or find medical, legal, or social support. Unmarried Syrian refugee women report feeling vulnerable as they have experienced sexual harassment and violence, as well as solicitations for sexual favors from strangers who knew about their non-marital status. Because of restrictions on their mobility, they find it impossible to report their situation to NGOs or to the authorities, and they fear that any attempt to report rape could result in their incarceration and deportation (Yasmine & Moughalian 2016).

Taking care of women's reproductive health

Reproductive and sexual health in Lebanon is dominated by the medical profession, and midwives are only legally allowed to perform their function as obstetricians, which means that the rate of hospitalization for the entire maternity process is extremely high. As mentioned previously, the healthcare system in the country is semi-private, and the state only provides medical protection to specific categories of *citizens*, meaning that Syrian refugees are excluded from any form of fee waiver when it comes to mental health. Lebanon lacks basic information on the organization, distribution, and quality of maternal health services, and hospital-level maternal and newborn health outcome data are neither publicly available nor systematically collected.

These deficiencies in the information base are due to the interrelated factors of the legacy of a 15-year civil war (1975–1990) and the limited role of the state in a health sector heavily dominated by the private sector (DeJong et al. 2010). In turn, the quasi-private nature of the system renders pregnancy and delivery another part of the profit-making process, which results in a high percentage of caesarean sections performed at about 40% of total deliveries, compared to the 15% recommended by the World Health Organization, indicative of a non-addressed problem of obstetric violence (DeJong et al. 2010). Obstetric violence is widespread among Syrian refugees, who report experiencing negligence, poor pain management, and lack of communication with healthcare providers during pregnancy and maternity. Refugee women are also exposed to higher caesarean rates (Yasmine & Moughalian 2016). Many migrant and refugee women are hesitant about delivering in hospitals for fear of having to pay for unnecessary tests and unreasonably high hospital bills, or being asked about the legal status of their residency and their sponsor.

These factors may prompt poor and vulnerable women to deliver at home. Given that maternal mortality rates are assessed through hospital registrations of maternal deaths, it is possible that the national rate does not adequately reflect the realities of poor and vulnerable women who are driven to home deliveries; the maternal death rates in the state may indeed be far higher, considering the number of deaths cases happened outside the hospitals (Yasmine & Moughalian 2016).

The UNHCR currently covers 75% of the cost of delivery for Syrian refugees; however, this still means that giving birth may come at an extremely high cost to the families involved. Many hospitals require the 25% upfront before the delivery and the subsequent procedures. At the other extreme, four doctors and aid workers interviewed by Refugees Deeply have testified that some Lebanese hospitals have confiscated identity documents or detained newborn babies until bills were paid – practices the Ministry of Health considers illegal (Karas 2017).

One-third of the total UNHCR budget in Lebanon goes in health expenditures due to the high prices of healthcare services in the country. In 2017, the UNHCR spent almost 50 million USD on healthcare, 74% of which was paid for referrals regarding women. This represents the highest-ever budget invested by the organization in healthcare provision in a single country (Karas 2017). Even with such high expenditures, the UNHCR funds and the funds provided by NGOs are insufficient to guarantee proper prenatal and post-delivery care; for example, Syrian refugees struggle to obtain the recommended ultrasound check-ups during pregnancy.

The absence of a legislative instrument to regulate abortion is another structural factor that influences the reproductive choices of both Lebanese and Syrian women. Abortion is simply illegal in Lebanon in all circumstances, except for the preservation of the woman's life when endangered by the pregnancy itself, as provisioned by Presidential Decree No. 13187 in 1969 (Fathallah 2019). In 2013, the Order of Pharmacists of Lebanon distributed a circular to all pharmacies instructing them to demand, examine, and keep any prescription for Cytotec (an abortifacient); abortion medications were previously much more accessible. With restrictions on access to abortion services impeding bodily autonomy, many women face compulsory motherhood (Rola & Batoul 2018).

Although the reproductive justice framework stresses that women should be able to make their reproductive choices with the availability of non-procreative sex without stigma, contraception, or abortion, yet none of these issues is stigma-free or free from sexist and incorrect beliefs. The main fault of unwanted pregnancies is always placed on men. Fathallah (2019) reported that when the abortion patient is an unmarried woman, many of the doctors usually shame her and question the irresponsibility of her choices, constraining such girls to justify their condition (Rola & Batoul 2018). Similarly, although contraception is not criminalized in the country, almost all the burden is placed on women and not on men, who are offered only two choices, condoms and vasectomies; the latter, even though they are reversible, are not often practiced because of the widespread fear of emasculation and possible sterility (Rola & Batoul 2018).

The Population Reference Bureau documents that the latest data (prior to 2006) in Lebanon showed that 58% of married women between the ages of 15 and 49 use some form of contraceptive method, with 34% of them using modern methods. This leaves the remaining 42% of those women with an unmet need for family planning resources, not to mention those engaging in premarital sexual activities (Fathallah 2019). Studies have shown that contraception was used among Syrian women (mostly those who had already had their desired number of children) in their homeland, mostly because it was free under the Family Planning Unit.

As such, women were able to consult a specialist and decide with the doctors on the best option for them. IUD and contraceptive injections every three months were the most popular options. Women displaced in Lebanon have complained not only about the unaffordability of these services, but also about their sheer lack.

Three different studies (Cherri et al. 2017; Spencer et al. 2015; Kabakian-Khasholian et al. 2017b) report the fact that in Lebanon, contraceptive injections, which were one of the most widespread and comfortable methods among Syrian women, are not available, and many women do not rely on the pill in the same way they relied on the injection since it is easier to make mistakes with them or forget to take one, resulting in the unstable situations in which some of these women live.

Micro-level analysis: personal relations

As in a dramatically high number of societies, the reason behind the reduced reproductive options given to women is men. Syria, prior to the crisis, was a patriarchal society. Despite the gains made by women in the 20 years prior to the uprisings, men still dominated almost all aspects of society, as both the holders of political office and the main breadwinners and protectors of the family (Kabakian-Khasholian et al. 2017b). The sexual life of the couple can be included in the chain of dominance. Men are reported to be among the main reasons for women not taking contraception (Spencer et al. 2015), while various articles report many women being dependent on their husbands' will around reproductive matters (Cherri et al. 2017; UNFPA 2018; Spencer et al. 2015; Karas 2017; Kabakian-Khasholian et al. 2017b). First, there is general misinformation about the effects of contraceptive measures, since many women refer to their partners' vetoes because of their fear of infertility, both when it comes to condoms for men and hormonal intakes for women. In a study led by Spencer et al. (2015), 28% of the women who participated in the study reported that their husbands physically impeded them from using contraception, either because they refused to give the money necessary to buy the products or by dispensing with the products themselves. A further 6% reported that this behavior was also imposed by family members, usually the in-laws:

> *If the woman does not want to use a [contraceptive method] that messes with her hormones, [she suggests the condom to her husband]. The final decision is taken by the husband. He either uses a condom or he does not.*
>
> (Cherri et al. 2017)

Another important push factor that prevents women from accessing contraceptive methods or abortions is the fear of polygamy, which is a constant threat. In all the studies cited earlier, women report the necessity of getting pregnant shortly after marriage, because waiting too long to generate offspring may compel the husband to search for another wife who will fulfil her reproductive duty:

> *If I disagree with him he will marry another woman, he will say I want children.*
>
> (Kabakian-Khasholian et al. 2017a: S81)

The prospect of men finding another woman, either for polygamous reasons or after divorce, is extremely scary for women due to their condition of being forcibly displaced. As we have already mentioned, marriage in displacement is a sort of guarantee for women from the economic point of view, because the husbands have to provide for them and their offspring, and from a protective point of view, since married women – compared with unmarried ones – are less targeted by locals and are no longer reliant on the protection of their fathers and brothers, who can be deceased because of war.

Of course, men also have an important influence in deciding the number of children, after which it is appropriate to stop reproducing. The male heir is important in determining the number of children; many couples who have many children usually do not stop until they have one, reaching numbers as high as 6–8 children. For those with male offspring, the number is influenced by societal patterns. In Syria, in recent decades, the number of children per family has decreased, from women who are now between 40 and 60 having 8–10 children to women now having an average of 3–4 children per family. The women interviewed in displaced contexts have expressed a grievance about the necessity of reducing the number of offspring because of their current situation (Cherri et al. 2017).

Changing perceptions and coping mechanisms

An article published by the *New Arab* in 2015 was titled Futures uncertain, Syrian refugees in Lebanon start family planning. The studies cited earlier confirm this new trend. Women have acquired new perspectives on their reproductive decisions and choices in displacement as well as new bargaining methods and coping mechanisms to engage with the hardships that come with living in Lebanon (Kabakian-Khasholian et al. 2017b). The women who participated in the interviews conducted by Kabakian-Khasholian expressed a desire to have between four and six children in the prewar context. However, they also stressed that these wishes have been scaled back after the war, especially after the worsening of their legal and living conditions in Lebanon.

Although the number of children has not diminished drastically, they have begun to have different needs in accessing family planning. The reasons behind wanting a large family are various. Some believed numerous offspring means being better attended when the parents become old; others observed that the initial UNHCR registration process and aid distribution were more favorable to families with more children, who were assisted with more food income and more possibilities of being resettled in a third country, especially in Europe (Kabakian-Khasholian et al. 2017a). A woman interviewed by Kabakian-Khasholian states as follows:

> Let's assume I got older; I have to make sure to have male children before getting old. The girls will get married and leave. I need to have at least one or two boys so that they support me when I get older . . . they can also help their sisters.
>
> (Kabakian-Khasholian et al. 2017a: S84)

However, with the prolongation of displacement, the outlook has changed: Syrians now face an ongoing process of shrinking livelihoods. The UNHCR and NGOs have facilitated Syrian women's access to healthcare, and while focusing on maternal health and offering aid to pregnant women and their families, they have played a significant role in determining women's willingness to have more children.

However, with the prolongation of the conflict, the UNHCR support has diminished, leaving the community with an even more accentuated experience of hardship, which is now influencing their decision to limit their family size. Consequently, personal preferences have been abandoned in favor of a more pragmatic approach. As one young Syrian woman reports:

> *One boy and one girl is more than enough in this situation, the fewer the better. I mean, life is difficult, everything is expensive, you need to be able to provide for both children and save money.*

This is a rising trend among younger women, especially those aged 18–25 (Kabakian-Khasholian et al. 2017a: S80). Women's main rationale for these reproductive changes is mostly economic in nature:

> *You have to pay for rent, electricity, water, and if the water is cut off, you need to buy it, everything changed, my husband needs to support everyone living with us here.*

Additionally, women face the prospect of giving birth to children who will have to access substandard education and with no certain future:

> *We have a big problem. I have three children and there is no school, and I cannot afford to send them to a private school. I teach them letters and how to count at home.*
>
> (Kabakian-Khasholian et al. 2017a: S80)

Some women cited their own health and personal lives as reasons for not having many children. Considering the current situation in Lebanon, the decreased number of children has started to be accepted by some husbands. As one interviewee explains:

> *The situation in Syria is different from here. In Syria, it is normal to give birth and [let the children grow] freely. The country is ours, and the situation is known there, but here no [it is different]. If anything [bad] happens to the parents here, what happens to the children? Hence, [having] fewer children is better.*
>
> (Cherri et al. 2017: 67)

Despite this, young women who have not achieved their desired number of children tend to avoid using contraception because of the fear of infertility or other consequences on their reproductive potential. Syrian women have expressed shared preoccupations over the consequences of hormonal intake, and many prefer natural methods of contraception, such as the rhythm method. Furthermore, the side effects of contraceptive pills, injections, and IUDs are also perceived as incompatible with poor living conditions, since they may cause additional illness in an already difficult health framework (Kabakian-Khasholian et al. 2017b).

Interestingly, economic hardship and the precarity of displacement, along with the shifting roles of men and women, are now being used as leverage by women to bargain their reproductive desires with their husbands. As we have explained earlier, husbands have a dominant position in the kinship, and women interviewed by the different studies have all explained that it was in their own interest to fulfil their family wishes because of the possible consequences (whether polygamy or divorce) that are considered to add liability to their position in the family.

In displacement, women who have to function as breadwinners and providers for their families have a new strategic position to discuss their new perspectives on fertility. Displacement has given men a different perspective on this issue, too:

> *I could not negotiate this issue with him before. If the man wants a child, you cannot prevent him, but now I have good reasons to convince him and to say no*
>
> (Kabakian-Khasholian et al. 2017b: 19).

Many women use the interplay of contextual factors to explain their will to the husbands, who, also in light of their shrinking livelihoods, are now more likely to listen and perhaps accept the women's propositions. In the case of married women who have many children and want to have abortions, physicians who are aware of the impact of socioeconomic conditions are often willing to terminate the pregnancy (Fathallah 2019). As such, refugees for whom abortion is economically viable may undergo the procedure. Due to the high risk for Lebanese specialists performing these procedures under the law, physicians usually end up charging clients for their work and for the risk of legal prosecution, which can lead to very high and unaffordable prices. With no regulation on the prices of abortion, clients are typically charged between $300 and $1,200, depending on the geographical location and the space in which the procedure is performed: home, clinic, or hospital (Kaddour et al. 2005).

The problem with accessing abortion for Syrian women, aside from abortion being one of the strongest social taboos, is the constant fear of being arrested and deported because they are undergoing a procedure that is considered illegal. Furthermore, although women may be prone to reducing their number of children, the majority of participants in the studies report that once a woman gets pregnant, there is no other morally possible alternative to keeping it:

> *No, it is forbidden [to abort]. When a woman is pregnant, that's it. There is nothing she could do.*
>
> (Cherri et al. 2017: 844)

This belief is accompanied by limited knowledge about emergency contraceptive methods among the refugee population (Kabakian-Khasholian et al. 2017b).

For those whose desires have not changed due to displacement, motherhood comes at a very high price, one that is unaffordable for most families due to the restrictions on labor and the lack of legal documentation. Indeed, even before the refugee influx, the Lebanese health system was not equally accessible to Lebanese citizens due to structural problems. For a vaginal birth (caesarian birth), a Syrian woman can be charged between 100 and 200 USD, plus the UNHCR waiver. Many are requested to deposit the sum before delivery.

Although the donations have had the positive effect of increasing the number of primary health clinics available to both refugees and Lebanese citizens, the healthcare system is overburdened by the presence of the refugees, and it is too underfunded to implement an effective response. For pregnant Syrian mothers, the only organization providing free deliveries is Médecins sans Frontières (MSF), which operates three mother-and-child health clinics in Lebanon: one in Beirut, one in the North, and one in the Bekaa Valley.

More than 300 babies are born at MSF clinics each month (Karas 2017). To respond to the health and maternity needs of Syrians, two phenomena are at stake: first, and unique in this context, is that women used to cross the border into Syria to access healthcare and maternity services. One midwife working at Médecins sans Frontières noted that the high cost of delivering in Lebanon has rendered it common for Syrian women to return to war-torn Syria to deliver for free (Yasmine & Moughalian 2016). This process was set in motion especially by women living close to the border between the two countries in the first years of the conflict, mostly because women and children were not initially targeted by the army; however, they soon began to be arrested to force their male family members to follow the forced conscription duty.

The second interesting coping mechanism is to ask for help from informal health workers who are not allowed to practice in the formal Lebanese healthcare system; to enter the official labor force, they would be required to follow the regular Labor Laws and thus find a sponsor and attain a residence permit. According to Honein-AbouHaidar et al. (2019), the Lebanese environment prevents these informal health workers from bringing relief to very complex circumstances by filling a gap in the formal health system and helping alleviate the burden of refugee health needs, as well as improving their own financial situation by charging extremely small amounts of money compared with the costs of the same operations in the formal health care system.

In the studies cited earlier (Kabakian-Khasholian et al. 2017b), Syrian women expressed a preference for female midwives, as they feel more at ease communicating with them about personal issues and undergoing vaginal exams. According to the study conducted by Ismail et al. (2018), a significant percentage of the health workers displaced in Lebanon are actually physicians or obstetricians and gynecologists, which could ease the situation for Syrian refugees not only economically but also in addressing their reproductive patterns with the greater insight that Lebanese doctors frequently lack, resulting in discriminatory and xenophobic attitudes toward Syrian women:

> *I am just like them – I am displaced, I am financially unstable. . . . I am dealing with the same problems. In addition to this, our traditions and culture are the same, which helps them relax more and feel more comfortable. Our pain is one.*
>
> (Ismail et al. 2018: 21)

In particular, women who come from rural areas, where families are more numerous, are facing growing prejudices when coming into contact with Lebanese doctors who consider them ignorant and underdeveloped and do not understand the notion of having many children, given the circumstances (Kabakian-Khasholian et al. 2017a). On the contrary, among Syrian providers, there is a higher level of trust because of cultural affiliation. As one Syrian health worker assesses:

> *Syrians here, especially those in the camps, are from the rural areas in Syria, so they are a bit difficult to interact with. Some of them cannot read or write. They don't know how to come to appointments or interact with the doctor, so when they were attending Lebanese clinics or seeing Lebanese doctors, they were exposed to maltreatment. We know how to deal with them because we used to live with them anyway.*
>
> (Ismail et al. 2018: 21)

Informal health workers are financially more accessible; some work for NGO clinics that provide services for free or apply a blanket payment of 2 USD for all services or only ask for what the patient can afford to pay. They are geographically more available than Lebanese doctors, who are mostly clustered in Beirut, and the road to the capital is full of checkpoints that stop Syrian refugees from going there (Ismail et al. 2018).

Outcomes and issues still under question

In the introduction, a series of questions was posed about the nature of the constraints and inducements that shape women's reproductive choices, especially if some of them derived from the policies enforced by the host state in which women are displaced. The principal outcome of this study is the fact that the utilization of the social ecological model has given us back the idea that, like violence against women in the study led by Heise (1998), violation of the reproductive rights and the right to parenthood of

Syrian refugees in Lebanon are not only connectable to individual wrongs or a cultural vacuum, but it is also perpetrated on many different levels, in between which women find spaces and measures to put forth their personal will, despite the extremely negative construction of "helpless" and "undeveloped" women that endures in the MENA region. Of course, it would be extremely unjust to say that women are always capable of transforming their wishes and desires into practice, in reproductive matters as well as in other matters, since many women suffer from a variety of unspeakable violence, which impose on them a never-ending series of burdens that displacement creates or accentuates. As we can see, society at large, from the micro level to the macro level, enables an oppressive environment in which Syrian women make or do not make decisions about their bodies.

According to the principal international human rights instruments and the provisions contained in them, women worldwide are entitled to have access to information about sexual and reproductive rights and health as well as access to services and medicine. They are entitled to give their consent, having been granted that their choices are made on the basis of reliable information, and when they choose to be mothers, they are to be protected in the name of the right to a safe and healthy pregnancy. In the words of Ross and Solinger (2017: 9), they are entitled to all the essential means that empower their right not to have a child, right to have a child, and their right to parent children in safe and healthy environments. In reality, however, Syrian women who are forcibly displaced in Lebanon are not granted these rights. Instead, they are pushed and pulled in different directions: by their partners, employers, extended families, and the Lebanese national authorities, which refuse to give them protection from any point of view.

The displacement experience for all Syrian nationals who have fled the conflict since 2011 has been steeped in xenophobia, sexism, racism, and despair. The Lebanese response throughout the influx has worsened as time passed, rendering the country inhospitable for those displaced within its borders. Their situation is extremely vulnerable because the host state cannot be defined as such in legal terms, since Lebanon refuses to ratify the Refugee Convention of 1951 and its Protocol, the specific political reasons for which are rooted in the precarious balance that holds the fragmented Lebanese society together, a society involved in an intensive civil war from the 1970s until the mid-1990s, and in which the historic refugee population of Lebanon, the Palestinians, also participated.

The Syrian situation is further complicated by the strong prejudices that are spread among the Lebanese, given that Syrian citizens, and often soldiers, occupied part of Lebanese territory until 2006. It is important to stress here that most Syrian men had no choice in participating in the Syrian occupation, given their government's oppressive, intransigent, and militarized stance when it comes to defining its source of power. Consequently, there is a general wish for the Syrians to return to their country, which is enforced by laws that violate human rights standards, but which have been approved and enforced, especially in recent years.

If the situation continues to degenerate for all Syrians in Lebanon, women must bring their reproductive choices to the forefront of the discussion. The general Lebanese xenophobic discourse, as we have seen in many of the studies collected for the purpose of this research, is that if Syrian refugees cannot afford to sustain the upbringing of so many children, they should not have as many as they do. However, this view lacks perspective. The total fertility rate has declined in Syria from a very high level of 5.3 in 1990 to 2.9 in 2010 before the conflict.

Nevertheless, in 2010, Syria had the sixth highest total fertility rate in the Arab world, and the average number of children was found to be higher in Syria than in Lebanon. Lebanese health providers harshly criticize Syrian women for the number of children they have, but they do not realize that families in Syria were more numerous thanks to the public nature of health-care, social welfare, and education, which allowed parents to have more children without worrying about providing their children with the basic standards of living, which is very far from the reality that Lebanese citizens experience given the dramatic social differences in their heavily privatized country.

Conclusions: the importance of the informal health care practices

Today in Lebanon, Syrian women are struggling to find free or affordable reproductive healthcare and are mistreated for being pregnant. A lower per-centage of women were found to be using family planning methods (34.5%) compared to before the conflict in Syria (58.3%). Before 2000, pro-natal-ist policies in Syria incentivized multi-parity, and family planning services continue to be free of charge and widely available. Conversely, the Leba-nese health sector has prohibitive costs when compared with the incomes of displaced Syrian families, the vast majority of which live in conditions of extreme poverty, according to the VASyR reports. In turn, most Syrian women have poor sexual and reproductive health. They do not visit gynecol-ogists even when pregnant because of fear, stigma, and lack of knowledge of reproductive health.

Most refugees also live outside Beirut, which means that they can access health clinics only by taking transportation that they cannot always afford. The same goes for contraceptives, which either they cannot afford or they cannot afford to implant, even when given for free.

This contributes to the high fecundity of the Syrian population. In the context of the exo-system, the Lebanese setting also prevents abortion if it is not to preserve the health of the mother at risk by the pregnancy itself. Com-bined with the poor socioeconomic conditions of Syrian refugees, abortions are not accessible for refugee women, since the price of an illegal abortion varies between 150 USD and 2,400 USD. Therefore, Lebanon, as a host country, fails as a country of reproductive justice and treats even its own women as second-class citizens.

For example, Lebanese women cannot pass their nationality on to their children if married to non-national men. Moreover, Lebanon fails to grant displaced Syrian women the right to parenthood, which, according to the reproductive justice framework, is necessary if women want to parent. Syrian children are not raised in a socially just or safe environment, especially girls.

From a micro-level point of view, in terms of personal history and relationships with partners, families and women are also subjected to inducements. Some of these derive from the patriarchal nature of Syrian society even before displacement; the vast majority of the women interviewed depend on their husbands' desires when it comes to having or not having a certain number of children or taking contraceptive measures, largely for fear of divorce or polygamy.

These constraints emanate from the religious background, although abortion, as in the Christian context, is highly debated, and from the religious and social construction of motherhood, perceived in both as the most important role preserved for women. Many of the women, although highly educated, were still married at a young age, also because of the unprotected status given to unmarried women.

The same thinking underpins the growing and disturbing practice of marrying girls off at a very young age, a phenomenon that was already present in rural Syrian society but which has been exacerbated by the hardship of displacement, especially because parents are not able to provide for their children and are in fear for their own lives, which would leave their daughters without protection. Young girls and unmarried girls are extremely vulnerable to sexual violence, harassment, and abuse, perpetrated by both co-nationals or the Lebanese, which puts their reproductive health and reproductive choices in an extremely precarious situation (Abdelwahed et al. 2018).

Many unmarried women are forced to engage in transitional sex to survive and make ends meet, and these women are continually subjected to shaming because of their survival activities. In this scenario, it is evident and clear that displacement has had some significant consequences for women and their reproductive patterns, especially those of younger married couples and younger women, since it has the capacity to produce a change in the socially constructed roles that men and women are expected to perform traditionally. Studies also show that the Syrian population has not been immune to these changes, especially because for many women breadwinning and providing for the family comes at a very high price, including increased domestic violence, intimate partner violence, or forced relegation to the domestic sphere as a means of restoring the status quo of pre-conflict society.

On the other hand, displacement has also had the positive effect of rendering partners more reasonable when presented with the economic and legal hardships of the Lebanese context as a motive to have a reduced number of children. In the past, when the Syrian health system was not unavailable as it is now, women displaced near the border used to return to Syria, especially if their children had postnatal issues, where they could be treated free of charge.

Now, this possibility is hardly available, and returns to Syria are very dangerous for women, so many of them are forced to rely completely on informal healthcare facilities, mostly put in place by NGOs, while others address their problems with the help of Syrian informal health workers, who are unauthorized to exercise their profession in Lebanon. These health workers represent an extremely valuable resource currently being wasted in the name of xenophobic principles, and in this the responsibility of the host environment is extremely important, considering its dramatic impact on those women and couples who have been willing to renounce the achievement of the desired family size simply because they have to cope with the shrinking livelihoods available for Syrian refugees in Lebanon. The embodied consequences of political decisions are dramatically tangible, no matter whether they are produced intentionally, as with the law forbidding free access to abortion, or unintentionally, by limiting the legalization of refugee status, which directly impacts the bodies of the women and their mindset in decision making over these bodies.

References

Abdelwahed, I., Abla, R., & Afifi, R. (2018). Recent news coverage of sexual and reproductive health in Lebanon. *Journal of Middle East Women's Studies, 14*(3), 390–393.

Abouelnaga, S. (2018). I am not a mother therefore I don't exist. *Kohl: A Journal for Body and Gender Research, 4*(2), 197–204.

Al Zoubi, S. T. (2019). Syrian refugees in Lebanon: Limited livelihoods and untold challenges. *University of Oxford. Refugee Studies Centre.* Retrieved from https://www.qeh.ox.ac.uk/sites/www.odid.ox.ac.uk/files/Workshop20booklet.pdf

Amnesty International. (2016). *Refugees are in urgent need of protection from sexual and gender-based violence.* Retrieved December 9, 2019, from www.amnesty.org/en/latest/news/2016/11/refugees-are-in-urgent-need-of-protection-from-sexual-and-gender-based-violence/

Amnesty USA. (2013). *The quick way you can take action for Syrian Women facing gender violence.* Retrieved December 9, 2019, from www.amnestyusa.org/the-quick-way-you-can-take-action-for-syrian-women-facing-gender-violence/

Atkinson, M., Greenstein, T., & Lang, M. (2005). For women, breadwinning can be dangerous: Gendered resource theory and wife abuse. *Journal of Marriage and Family, 67*(5), 1137–1148.

Avis, W. R. (2017). *Helpdesk report: Gender equality and women's empowerment in Lebanon.* Retrieved April 24, 2020, from https://reliefweb.int/report/lebanon/helpdesk-report-gender-equality-and-women-s-empowerment-lebanon

Badissy, M. (2016). Motherhood in the Islamic tradition rethinking the procreative function of women in Islam. *Muslim World Journal of Human Rights*, 13(1).

Bashour, H., & Abdulsalam, A. (2005). Syrian women's preferences for birth attendant and birth place. *Birth*, 32(1), 20–26.

Bilgili, O., Loschmann, C., & Siegel, M. (2017). The gender-based effects of displacement: The case of Congolese refugees in Rwanda. KNOMAD Working Paper, 21.

Bradley, S. (2018). Domestic and family violence in post-conflict communities: International Human Rights Law and the State's Obligation to Protect Women and Children. *Health and Human Rights*, 20(2), 123–136.

Brush & Bow. (2019). Podcast #8 Randa: Exploitation in Lebanon. *Radio HAKAYA*. Retrieved November 15, 2019, from https://brushandbow.com/radio-hakaya-%d8%ad%d9%83%d8%a7%d9%8a%d8%a7/

Chehayeb, K. (2019). Lebanon troops demolish Syrian refugee homes as deadline expires. *Al Jazeera*. Retrieved from https://www.aljazeera.com/news/2019/7/1/lebanon-troops-demolish-syrian-refugee-homes-as-deadline-expires [accessed 17 May 2022].

Cherri, Z., Gil Cuesta, J., Rodriguez-Llanes, J., & Guha-Sapir, D. (2017). Early marriage and barriers to contraception among Syrian Refugee Women in Lebanon: A qualitative study. *International Journal of Environmental Research and Public Health*, 14(8), 836.

DeJong, J., Akik, C., El Kak, F., Osman, H., & El-Jardali, F. (2010). The safety and quality of childbirth in the context of health systems: Mapping maternal health provision in Lebanon. *Midwifery*, 26(5), 549–557.

El Arnaout, N., Rutherford, S., Zreik, T., Nabulsi, D., Yassin, N., & Saleh, S. (2019). Assessment of the health needs of Syrian refugees in Lebanon and Syria's neighbouring countries. *Conflict and Health*, 13(31).

Fathallah, Z. (2019). Moral work and the construction of abortion networks: Women's access to safe abortion in Lebanon. *Health and Human Rights Journal*. Retrieved January 20, 2020, from www.hhrjournal.org/2019/12/moral-work-and-the-construction-of-abortion-networks-womens-access-to-safe-abortion-in-lebanon/

Freedman, J. (2015). *Gendering the international asylum and refugee debate* (2nd ed.). Basingstoke, UK & New York: Palgrave Macmillan.

Freedman, J. (2017). Women's experience of forced migration: Gender-based forms of insecurity and the uses of "vulnerability". In J. Freedman, Z. Kivilcim, & N. Özgür Baklacıoğlu (Eds.), *A gendered approach to the Syrian refugee crisis* (pp. 125–142). Oxford and New York, NY: Routledge.

Geha, C. (2019). Nobody knows Lebanon's problems better than its women. It's time you started listening. *The New Arab*. Retrieved January 17, 2020, from www.alaraby.co.uk/english/comment/2019/11/8/nobody-knows-lebanons-problems-better-than-its-women

Greene, M., & Merrick, T. (2006). *Poverty reduction: Does reproductive health matters?* Retrieved January 17, 2020, from http://siteresources.worldbank.org

Habbal, T. (2019). Syrians in Lebanon: A life of misery, or a return to the unknown. *Atlantic Council*. Retrieved November 15, 2019, from https://www.atlanticcouncil.org/blogs/menasource/syrians-in-lebanon-a-life-of-misery-or-a-return-to-the-unknown/

Heise, L. L. (1998). Violence against women: An integrated, ecological framework. *Violence Against Women*, 4(3), 262–290.

Hessini, L. (2007). Abortion and Islam: Policies and practice in the Middle East and North Africa. *Reproductive Health Matters, 15*(29), 75–84.

Hidayatullah, A. A. (2014). Feminist interpretation of the Qur'an in a comparative feminist setting. *Journal of Feminist Studies in Religion, 30*(2), 115.

Honein-AbouHaidar, G., Noubani, A., El Arnaout, N., Ismail, S., Nimer, H., Menassa, M., & Fouad, F. M. (2019). Informal healthcare provision in Lebanon: An adaptive mechanism among displaced Syrian health professionals in a protracted crisis. *Conflict and Health, 13*(1).

Human Rights Watch. (2018a). *Lebanon: Discriminatory nationality law.* Retrieved January 17, 2020, from www.hrw.org/news/2018/10/03/lebanon-discriminatory-nationality-law

Human Rights Watch. (2018b). *Lebanon: Mass evictions of Syrian refugees. Expulsions by municipalities appear discriminatory, Lack due process.* Retrieved November 15, 2019, from www.hrw.org/sites/default/files/report_pdf/lebanon0418_web.pdf

Hynes, M., & Cardozo, B. L. (2000). Observations from the CDC: Sexual violence against refugee women. *Journal of Women's Health & Gender-Based Medicine, 9*(8), 819–823.

Içduygu, A., & Nimer, M. (2019). The politics of return: Exploring the future of Syrian refugees in Jordan, Lebanon and Turkey. *Third World Quarterly,* 1–19.

Ismail, S. A., Coutts, A. P., Rayes, D., Roborgh, S., Abbara, A., Orcutt, M., Fouad, F., Honein, G., El Arnaout, N., Noubani, A., Nimer, H., & Rutherford, S. (2018). *Refugees, healthcare and crises: Informal Syrian health workers in Lebanon.* London: IIED. Retrieved from http://pubs.iied.org10856IIED

Janmyr, M. (2018). UNHCR and the Syrian refugee response: Negotiating status and registration in Lebanon. *The International Journal of Human Rights, 22*(3), 393–419.

Kabakian-Khasholian, T., Mourtada, R., Bashour, H., El Kak, F., & Zurayk, H. (2017a). Perspectives of displaced Syrian women and service providers on fertility behaviour and available services in West Bekaa, Lebanon. *Reproductive Health Matters, 25*(Suppl. 1), S75–S86.

Kabakian-Khasholian, T., Bashour, H., El-Nemer, A., Kharouf, M., Sheikha, S., El Lakany, N., Barakat, R., Elsheikh, O., Nameh, N., Chahine, R., & Portela, A. (2017b). Women's satisfaction and perception of control in childbirth in three Arab countries. *Reproductive Health Matters, 25*(Suppl. 1), S16–S26.

Kaddour, A., Hafez, R., & Zurayk, H. (2005). Women's perceptions of reproductive health in three communities around Beirut, Lebanon. *Reproductive Health Matters, 13,* 34–42.

Karas, T. (2017). For refugees in Lebanon, giving birth comes at a high price. *Refugees Deeply.* Retrieved January 20, 2020, from www.newsdeeply.com/refugees/articles/2017/07/07/for-refugees-in-lebanon-giving-birth-comes-at-a-high-price

Karasapan, O., & Shah, S. (2021). *Why Syrian refugees in Lebanon are a crisis within a crisis.* Retrieved April 28, 2022, from https://www.brookings.edu/blog/future-development/2021/04/15/why-syrian-refugees-in-lebanon-are-a-crisis-within-a-crisis/

Khodr, Z. (2019). Lebanon asks Syrian refugees to demolish their houses. *Al Jazeera.* Retrieved November 14, 2019, from https://www.aljazeera.com/news/2019/06/lebanon-asks-syrian-refugees-demolish-houses-190613120214474.html

Kranz, M. (2018). Lebanese security forces crackdown on Syrians as pressure builds on refugees to return. *The New Arab*. Retrieved November 15, 2019, from: https://www.alaraby.co.uk/english/indepth/2018/11/21/lebanese-security-forces-crackdown-on-syrian-refugees

Mhaissen, R., & Hodges, E. (2019). Unpacking return: Syrian refugees' conditions and concerns. *SAWA for Development and Aid*. Retrieved from https://reliefweb.int/sites/reliefweb.int/files/resources/SAWA_Unpacking%20Ret urn%20Report.pdf

Minority Rights Group International. (2019). *An uncertain future for Syrian refugees in Lebanon: The challenges of life in exile and the barriers to return*. Retrieved January 20, 2020, from https://minorityrights.org/wp-content/uploads/2019/02/MRG_Brief_Leb_ENG_Feb19.pdf

Muftić, L. R., & Cruze, J. R. (2014). The laws have changed, but what about the police? Policing domestic violence in Bosnia and Herzegovina. *Violence Against Women, 20*(6), 695–715.

Norwegian Refugee Council. (2014). *The consequences of limited legal status for Syrian refugees in Lebanon*. Retrieved November 13, 2019, from www.nrc.no/globalassets/pdf/reports/the-consequences-of-limited-legal-status-for-syrian-refugees-in-lebanon.pdf

Operazione Colomba. (2019). *Report Mensile, Agosto 2019*. Retrieved December 10, 2019, from https://www.operazionecolomba.it/dove-siamo/libano-siria/libanosiria-report/3267-agosto-2019.html

Pittaway, E., & Bartolomei, L. (2018). *From rhetoric to reality: Achieving gender equality for refugee women and girls*. Center for International Governance Innovation, Waterloo, Canada. World Refugee Council Research Paper No. 3.

Reuters. (2019). *Lebanon working for return of thousands of Syrian refugees: Security official*. Retrieved November 17, 2019, from https://www.reuters.com/article/us-mideast-crisis-syria-lebanon-refugees/lebanon-working-for-return-of-thousands-of-syrian-refugees-security-official-idUSKCN1IW26R

Robbers, G., Lazdane, G., & Dinesh, S. (2016). Sexual violence against refugee women on the move to and within Europe. *WHO Regional Office for Europe, 84*, 26–29.

Rola, Y., & Batoul, S. (2018). In the pursuit of reproductive justice in Lebanon. *Kohl: A Journal for Body and Gender Research, 4*(2). Retrieved January 17, 2020, from https://kohljournal.press/pursuit-rj-lebanon

Ross, L. J., & Solinger, R. (2017). *Reproductive justice: An introduction*. Oakland, CA: University of California Press.

Salameh, R. (2013). Gender politics in Lebanon and the limits of legal reformism. *Civil Society Knowledge Centre*. Retrieved January 17, 2020, from http://civilsociety-centre.org/paper/gender-politics-lebanon-and-limits-legalreformism-en-ar

Salloum, J., & Hodges, E. (2019). Unpacking return: Syrian refugees' conditions and concerns. *SAWA for Development and Aid*. Retrieved from https://reliefweb.int/sites/reliefweb.int/files/resources/SAWA_Unpacking%20Ret urn%20Report.pdf

Spencer, R. A., Usta, J., Essaid, A., Shukri, S., El-Gharaibeh, Y., Abu-Taleb, H., Awwad, N., Nsour, H., Alianza por la Solidaridad, United Nations Population Fund-Lebanon & Clark, C. J. (2015). Gender based violence against women and girls displaced by the Syrian conflict in South Lebanon and North Jordan: Scope of violence and health correlates. Alianza por la Solidaridad. Retrieved January 23, 2020, from www.alianzaporlasolidaridad.org/wp-content/uploads/GBV-Against-Womenand-Girl-Syrian-Refugees-in-Lebanon-and-Jordan-FINAL.pdf

UNFPA – United Nations Population Fund. (2018). *Among Syrian refugees, dispelling myths about contraceptives*. Retrieved January 25, 2020, from: https://www.unfpa.org/news/among-syrian-refugees-dispelling-myths-about-contraceptives

United Nations High Commissioner for Refugees, United Nations Children's Fund, & World Food Programme. (2018). *Vulnerability assessment of Syrian Refugees in Lebanon (VaSyr) 2018*. Retrieved from https://www.unhcr.org/lb/wp-content/uploads/sites/16/2018/12/VASyR-2018.pdf

United Nations High Commissioner for Refugees, United Nations Children's Fund, & World Food Programme. (2019). *Vulnerability assessment of Syrian Refugees in Lebanon (VaSyr) 2018*. Retrieved from www.unhcr.org/lb/wp-content/uploads/sites/16/2018/12/VASyR-2018.pdf

World Health Organization. (2017). *Global Health Observatory on maternal and reproductive health*. Retrieved January 3, 2020, from www.who.int/gho/maternal_health/en/

Yasmine, R., & Moughalian, C. (2016). Systemic violence against Syrian refugee women and the myth of effective intrapersonal interventions. *Reproductive Health Matters, 24*(47), 27–35.

5 LGBT activism in repressive contexts[1]

The struggle for (in)visibility in Egypt, Tunisia, and Turkey

with Aurora Perego

Drawing on social movements and gender studies, this chapter aims to explore the processes of mobilization and the survival strategies articulated by LGBT communities in Egypt, Tunisia, and Turkey during and after the 2011 and 2013 protests. The aim is to disentangle how LGBT individuals mobilized in the MENA region and what role civil society organizations and digital technologies played in the development of such mobilizations. The state repression of mobilizing structures and the relevance of digital networks in mobilization processes involving LGBT activists and individuals in the three countries will be discussed. The empirical analysis draws on 44 semi-structured interviews carried out in Egypt, Tunisia, and Turkey between 2011 and 2020, focusing on repressive contexts, civil society activism, and digital networks. By doing so, the analysis also aims to shed light on the roles played by both the more structured meso-level organizations and the more spontaneous digital technologies in triggering a range of diverse survival strategies.

In these three countries, LGBT communities have been disproportionately targeted by state and non-state repressive campaigns. In Egypt, LGBT activists have challenged repression through the use of social networks as alternative venues for socialization; in Tunisia and Turkey, LGBT activists have drawn upon more established meso-level mobilizing structures to create and implement new strategies, thereby increasing their cooperation with other political challengers.

Introduction

LGBT[2] activism in contexts where homosexual, bisexual, and transgender subjectivities are repressed and persecuted has increasingly gained academic attention. The literature indicates that LGBT individuals living in hostile environments face both state repression (Davenport 2007; Tschantret 2020) and soft repression (Ferree 2004). The former occurs through the (actual or threatened) use of physical sanctions (Davenport 2007: 2), for instance, via criminalization of same-sex relationships, police raids, or the death penalty. The latter concerns actions undertaken by non-state actors with the

DOI: 10.4324/9781003293354-6

aim of ridiculing and silencing marginalized communities (Ferree 2004: 88), such as media misrepresentation and the public delegitimization of their rights. Scholars have also examined how LGBT activism may develop under repressive conditions, why actors may participate in public mobilizations, and how their repertoires of action may evolve. However, these studies have mainly focused on the historical roots of western LGBT mobilizations since the Stonewall riots (Bernstein 1997; D'Emilio 1983) or on post-communist countries and Eastern Europe (Ayoub 2016; Buyantueva & Shevtsova 2019; O'Dwyer 2018). Except for a few investigations (e.g., Birdal 2020 and Fortier 2019[3]), research has not paid attention to how LGBT activism in the MENA region has developed since the Arab Spring and the Gezi Park movement, a period of comprehensive change for civil society in the region.

Drawing on social movement and gender studies scholarship, especially in the MENA region, this chapter explores how the repertoires of action (Tilly 1986) articulated by LGBT communities in Egypt, Tunisia, and Turkey have changed since the outbreak of the 2011 and 2013 uprisings. It addresses the following research questions: how have LGBT individuals based in the MENA region mobilized since the 2011 and 2013 protests? What role has repression against LGBT communities played in the development of such repertoires of action? How have LGBT civil society organizations (CSOs) and information and communication technologies (ICTs) contributed to the development of LGBT mobilization strategies in different repressive contexts?

This chapter investigates the nexus between the repertoires of action articulated by LGBT activists and the repression of LGBT communities in these three countries since 2011. To do so, it considers the role played by CSOs and digital networks in the evolution of the repertoires of contention developed by LGBT activists living under repressive conditions. The empirical analysis draws on 44 semi-structured interviews conducted by the authors in Egypt, Tunisia, and Turkey between 2011 and 2020 (Acconcia 2018).

LGBT repertoires of action in repressive contexts: the struggle for (in)visibility

Social movement scholars have long examined the nexus between repressive contexts and challengers' repertoires of actions, which are defined as the sets of means developed by challengers to make political claims (Tilly 1986: 2, 2008: 14). To explain how and why activists may articulate different repertoires of action, scholars have considered both macro-level conditions, such as the specific features of repressive contexts, and meso-level factors, such as the presence of CSOs and their use of digital media, which may strengthen challengers' capacity to innovate their repertoires of action. The role played by repressive conditions in activists' repertoires of action has been analyzed through the contrasting hypotheses of radicalization and moderation of the repertoires of action[4] (Davenport 2005; Earl

2003; Pilati 2016). However, scholars have also found that CSOs[5] play a crucial role in the innovation of activists' repertoires of action in repressive contexts (Pilati 2016) because they provide venues for political socialization (Tétreault 2000; Dorsey 2012). They may indeed ally with the central government (Jamal 2007), renovate their repertoires of action through the two main mechanisms of radicalization (Beinin & Vairel 2011) and moderation (Duboc 2011), or even opt for a depoliticization of their agendas (Bayat 2002; Clark 2004; Dorsey 2012). Furthermore, scholars have found that ICTs also contribute to the innovation of challengers' repertoires of action[6] (Rasler 2016). ICTs may enhance the radicalization of the repertoires of action by reducing the costs of information exchange, recruitment, and coordination (Hamanaka 2020; Howard & Hussain 2013) by reinforcing people's expectations for success, thus encouraging individuals to take political actions (Howard & Hussain 2011), and by helping activists mobilize public outrage (Hassenpour 2014; van de Bildt 2015). New technologies may also support the development of moderate and apolitical actions through the development of informal networks across different individuals and social groups (Rasler 2016).

Studies on LGBT activism in repressive contexts suggest that sexual minorities' repertoires of action stem from a complex tension between the need for public recognition and the risks of 'coming out' and being visible (Ayoub 2016; Birdal 2020; Fortier 2015, 2019; Wilkinson 2020). On the one hand, LGBT activists may seek public visibility with the aim of making their claims more resonant in the public sphere[7] (Currier 2012; Zivi 2012). Within this perspective, 'coming out' is considered a political strategy aimed at moving sexual rights from the margins to the center of political debates and achieving their full recognition (Ayoub 2016). The rationale of this tactic is that by claiming public space, non-heterosexual and non-cisgender identities will gradually be conceived as 'normal' instead of 'deviant' and will therefore be granted the same rights enjoyed by the rest of the population (Waaldjik 1994; Wilkinson 2020). On the other hand, in repressive contexts, higher visibility has often been followed by increased vulnerability and violence (Edenborg 2017, 2020; Wilkinson 2017). This situation, known as 'hypervisibility' (Wilkinson 2020), occurs when public authorities ally with anti-LGBT actors to raise concerns over non-heterosexual and non-cisgender identities and practices, for instance, by portraying homosexuality as a threat to the survival of the nation (Wilkinson 2014).

Revolutionary uprisings against authoritarian governments are crucial to the articulation of tactical repertoires (Almeida 2003; Davenport 2005; Goldstone & Tilly 2001). However, while uprisings against authoritarian regimes may empower certain sectors of the population, they usually endanger sexual minorities (Tschantret 2020). Transitional governments are indeed found to disproportionately target LGBT communities for both strategic and ideological reasons: on the one hand, political elites instrumentally repress less visible groups to show their ability to contrast national instability; on the other hand, sexual minorities may be perceived to be influenced by western-centric and liberal principles that may threaten the project of a new nation state (Tschantret 2020). Evidence of this phenomenon can also be found in the MENA region. LGBT individuals took an active part in the recent uprisings in the MENA region in the hope of gaining civil rights and social justice for sexual minorities in their respective countries (Birdal 2020; El Amrani 2019).

However, as soon as transitional governments were formed, LGBT individuals were subjected to repressive acts, such as police raids, illegal imprisonment, and public shaming (El Amrani 2019; Fortier 2015, 2019; Needham 2013). Despite gaining public visibility during the Arab Spring and Gezi Park movements, these individuals were forced into invisibility once the protests terminated.

Diversifying strategies to be (in)visible: organizational structures and digital networks

To address the struggle for (in)visibility in the post-revolt phase, LGBT activists in the MENA region have articulated various political strategies. On the one hand, they have aimed at increasing their visibility in the public sphere, with tactics ranging from creating stronger communities with the aim of triggering cultural change to building broad coalitions with the aim of campaigning for legal reforms and publicly addressing homophobia (El Amrani 2019). Such repertoires of action comprised participation in public demonstrations, the use of recognizable symbols like the rainbow flag, and the creation of coalitions with human rights organizations and networks (Fortier 2015). On the other hand, due to the increased violence against LGBT communities by postrevolutionary governments, LGBT challengers have also utilized strategies aimed at achieving higher discretion (Fortier 2015, 2019). Their repertoires encompassed the organization of informal meetings in cafes and universities (Fortier 2015), and the use of dating apps (Alqaisiya 2020). Building on the insights discussed earlier, we argue that LGBT activists in the MENA region may not only adopt different strategies to make their claims more or less visible in the public sphere, depending on the repressive conditions following the 2011 uprisings, but also according to the presence of CSOs and their use of social media.

In contrast to Egypt, where LGBT organizations were neither present before nor founded after the uprisings, LGBT groups were active in certain countries in the MENA region even before the Arab Spring (Birdal 2020). For instance, the first Turkish and Lebanese LGBT organizations were officially registered between the 1990s and 2000s. Other countries, such as Tunisia and Morocco, took advantage of the 2011 uprisings to establish their first organizations (Birdal 2020; El Amrani 2019; Fortier 2015, 2019). These organizations and groups have articulated different repertoires of action: on the one hand, they have increased the public visibility of LGBT claims and advocated for LGBT recognition and rights (Fortier 2015, 2019); on the other hand, they have acted as venues to support each other and create a safe community (El Amrani 2019). In a different fashion, some years before the outbreak of the uprisings, Egyptian LGBT activists had already been found to articulate an 'activism from the closet' strategy (El Menyawi 2006).

Rather than primarily advocating for LGBT rights in a society that did not perceive such claims as legitimate, these activists framed LGBT rights as part of a broader range of issues that concerned every citizen, such as human rights and freedom (Magued 2021; Needham 2013; Birdal 2020). By not directly addressing LGBT issues, this strategy was aimed at protecting activists' safety (Birdal 2020). Within this framework, LGBT individuals need not disclose their LGBT identities in the public sphere to engage in activism. On the contrary, the 'closet' was perceived as "a safe locus for collective strategizing" (El Menyawi 2006: 51).

The few studies on LGBT mobilizations in the MENA region show that activists have relied heavily on online platforms in the aftermath of the Arab Spring (Birdal 2020; El Amrani 2019; Needham 2013). Consistent with the struggle for (in)visibility discussed earlier, LGBT organizations have deployed digital platforms in different ways. On the one hand, the Internet has been used as a tool to "come out of the digital closet" (Gorkemli 2012). In other words, social media have served both to provide information about the aims and activities of the organizations and to publicly defy homotransphobic statements, promote petitions, advocate for LGBT rights, and denounce human rights violations (Fortier 2015, 2019).

On the other hand, with the increase of public repression of LGBT activists, Tunisian organizations have gradually shifted from the visibility strategy to the 'activism from the closet' approach (Fortier 2015). In a similar vein, Turkish LGBT activists and organizations deployed the Internet as a 'digital closet' during the 1990s (Gorkemli 2012). Indeed, the Internet afforded otherwise isolated individuals the possibility of gathering without 'coming out' in public. Moreover, during the early 2000s, Turkish LGBT organizations started using the Internet to organize widespread campaigns to make LGBT individuals and claims more visible (Gorkemli 2012). Hence, Turkish LGBT activists deployed digital platforms in a twofold way: first, to exchange information and create a community in which everyone could feel safe by not being exposed, and second, to defy mainstream negative representations of LGBT individuals by promoting coming-out strategies.

Within this framework, LGBT organizations and groups in the MENA region may be considered platforms for negotiating the struggle for (in)visibility articulated by activists (Sherif 2020). In other words, we expect that, depending on the configuration of repression against LGBT minorities in Turkey, Tunisia, and Egypt, LGBT organizations and groups may constitute both structures for visible advocacy and venues for invisible and safer community building. LGBT constituencies may negotiate their struggle for (in)visibility either by existing as networking venues for LGBT individuals living in hostile contexts or by resisting repressive conditions through advocacy and mobilization. Furthermore, they may deploy new technologies to negotiate the tension between the will to be visible and advocate for their rights in the public sphere, and the need to create an invisible community to protect their safety. Against this backdrop, digital media may be used as venues to exist as LGBT individuals by providing the means to communicate and exchange information with peers, as well as the means to resist under repressive circumstances by endowing challengers with platforms to connect and mobilize.

The empirical study

Case studies

To investigate the repertoires of action of LGBT communities in the MENA region, we focused on Egypt, Tunisia, and Turkey as cases of rather hostile contexts. Although Turkey has sometimes been excluded from some MENA studies, it has a growing and leading role in the region.[8] Moreover, the three countries are characterized by Sunni Muslim majorities and have witnessed historical tolerance toward local LGBT communities (Massad 2002) compared to neighboring countries, as well as repression after major episodes of mobilization, as seen in Egypt before and after the 2011 uprisings, in Tunisia before and in the aftermath of the 2010–2011 uprisings, and in Turkey before and after the Gezi Park movement (2013).

Data sources

The fieldwork research comprised 44 semi-structured interviews undertaken by the first author and the third author. The interviews were conducted with male and female Egyptian, Tunisian, and Turkish LGBT activists involved in grassroots mobilizations and advocacy campaigns, as well as ordinary citizens in Cairo, Alexandria, Tunis, Sousse, and Istanbul between 2011 and 2014 in Egypt and 2019–2020 in Tunisia and Turkey. Some of the interviews were conducted through several collective discussions. The testimonies offered insights and perspectives on the pre-, during, and post-2011 uprisings in urban and peripheral Egyptian and Tunisian neighborhoods and on mobilizations in university campuses before and after the Gezi Park movement (2013) in Turkey.

For the Egyptian interviewees (28), after an initial meeting with Revolutionary Socialist activists, a left-wing political group advocating for social rights located in Cairo, a snowball method was utilized to involve other participants. Thus, the selection of the interviewees started with contacts from founding members active in the Tahrir Square demonstrations and included additional participants via chain referrals to select both activists and ordinary citizens. In addition, gatekeepers working as NGO activists were interviewed in Cairo and Alexandria, and they also took part in the composition and organization of the interviews.

For the Tunisian interviewees (10), after a first meeting with Shams Association activists, a thinktank campaign for the depenalization of homosexuality in Tunisia, and individual supporters of the LGBT movements in the country, a snowball method was utilized to involve other participants. Thus, the selection of the interviewees encompassed contacts from initial members active in the Tunis and Sousse anti-regime protests in 2010–2011, as well as other participants comprising both activists and ordinary citizens.

The interviews in Turkey (6) involved supporters of the major local LGBT associations (Legato, Kaos-Gl, and Lambda) active in Istanbul and Ankara and within university campuses. Thus, the selection of the interviewees was based on contacts with LGBT individuals active in student associations, the Gezi Park movement (2013), and the 2015 Gay Pride to interview both activists and ordinary citizens.

The interviews were organized with the specific aim of understanding a range of topics: the involvement of Egyptian, Tunisian, and Turkish LGBT communities in grassroots mobilizations, their repertoires of action, police and military repression, media stigmatization, cooperation with other oppositional groups, mobilizations within campuses,[9] narratives of the 2011 uprisings and the Gezi Park movement (2013) and their respective aftermaths, and relations with state agencies, political parties, and Islamist groups. The interviewees appeared to be supportive of the Tahrir Square, Habib Bourguiba 2010–2011 demonstrations' demands as well as the Gezi Park movement (2013), participating in public protests in Egypt between January 2011 and June 2011, in mobilizations in Tunisia between December 2010 and January 2011, and in protests in Turkey before and after 2013. The interviewed LGBT supporters had been part of several waves of protests before the 2011 uprisings. However, in many cases, this participation had not been formalized and remained at the individual level.

Access to the field was very problematic, especially as a consequence of the increasingly repressive measures taken after the 2013 military coup in Egypt, under Article 230 of the Penal Code that criminalizes homosexuality in Tunisia, and the increasing repression after the 2016 failed coup in Turkey. The interviewees expressed security concerns with reference to their participation in the interviews. Consequently, all interviewees were anonymized, and each interviewee was assigned an identification number.

LGBT mobilizations in Egypt, Tunisia, and Turkey: (in)visible advocacy

Concerning LGBT mobilization strategies, the interviews show that LGBT individuals participated in protests taking place in Egypt, Tunisia, and Turkey with the aim of both supporting transformative claims and advancing

LGBT rights in the public sphere. However, these repertoires developed in different ways, both during and after the riots. While Egyptian activists tended to mobilize through informal gatherings and social media, Tunisian and Turkish activists also founded structured organizations to enhance advocacy strategies.

In Egypt, groups of young homosexuals used to gather in downtown Cairo long before the 2011 uprisings:

> *We used to meet there during the anti-Mubarak demonstrations [2006] and later on the occasion of the anti-Morsi protests [2013], or even on ordinary working days,*[10]

an Egyptian LGBT activist stated. According to our interviewees, LGBT supporters were among the various marginalized groups that gathered in Tahrir Square in 2011, hoping to see their rights recognized in the near future:

> *At the crossing between Tahrir Square and Talaat Harb Street, at the corner with the metro station Anwar al-Sadat, and on the metro gates in front of the KFC restaurant in Tahrir Square, another silent revolution was taking place,*[11]

an Egyptian LGBT activist taking part in the Tahrir Square protests explains. However, precisely because of their sexual orientation and gender identity, their participation in the riots made them, together with female activists (Chafai 2020; El Ashmawy 2017), specific targets of state repression:

> *In parallel with the repression of women participating in the 2011 protests, we faced the same kind of attacks, arrests, harassments, and anal probes,*[12]

another Egyptian LGBT activist outlined.

After the 2011 uprisings, meetings organized by LGBT activists continued in different parts of downtown Cairo for months, but without reaching their goals:

> *In 2011 and 2012, we were planning to stage a 'Cairo Gay Pride' in Tahrir Square many times, but only a few people joined us.*[13]

In this phase, social media and new technologies were crucial to LGBT mobilization. As an LGBT activist who took part in the Tahrir Square protests emphasized,

> *Thanks to the use of social networks, it was easier to bypass state control and organize protests for dissent.*[14]

Due to state repression, Egyptian LGBT mobilizations in the aftermath of the revolution were therefore still characterized by the presence of informal groups organizing meetings and gatherings through digital communication platforms.

If women were central to the success of the Tunisian 2010–2011 protests, there was widespread participation among LGBT activists as well. As one LGBT activist who took part in the Tunis demonstrations highlighted,

> *Many LGBT activists took part in the 2010–2011 protests at the individual level.*[15]

In contrast to Egypt, Tunisian activists affirmed that there had been a political opening following the riots. As an activist who took part in protests in Tunis explained,

> *In the aftermath of the uprisings, a wider space for LGBT activism was evident.*[16]

A positive consequence of such an opening may be seen in the establishment of Shams (Sun), Maghudin (We exist), and Chouf, the first three Tunisian LGBT organizations founded between 2012 and 2016. Shams was legally registered on May 18, 2015. However, these organizations were opposed by the central government, which presented a lawsuit against Shams to suspend its activities for one month starting from January 4, 2016.

Thus, Tunisian LGBT activists took advantage of a slight opening in the political opportunity structures to create more structured organizations from which to articulate public advocacy strategies. As an LGBT activist explained, this mobilization process has been slow but consistent over the years since:

> *There is not a structured organization or a political party advocating for LGBT rights, but a tradition of activism growing little by little since 2007. There is a new generation of activists, like Saif Ayadi, both LGBT and feminist militants, engaged in social struggles as well. This introduced a very interesting new dimension in the way in which political demands are formulated in Tunisia.*[17]

A similar mobilization process has been taking place in Turkey after the rise to power of the current president, Recep Tayyip Erdoğan, and the Islamic conservative Justice and Development Party (AKP) after 2003 (Yeşilada 2015). Feminists, LGBT activists, human rights groups, academics, nationalists, liberals, environmentalists, students, Kurds, and anti-capitalist Muslims protested against the Turkish authorities in 2013 during the Gezi Park movement.[18] As one LGBT activist explains,

> *This has been a chance for the LGBT movement to ask for a better inclusion of sexual minorities into society, while repressive government repeatedly and arbitrarily denied their rights to peaceful assembly.*[19]

Due to limitations on public gatherings between 2013 and 2020, LGBT associations based on university campuses have gained increased visibility. Legato (Lesbian and Gay Inter-University Organization) is a solidarity network among university students aimed at connecting LGBT people on campus.

Despite facing some problems with the university's internal authorities, the network members began to reorganize, contacting other universities in Istanbul and in other cities. Pembe Hayat (Pink Life LGBT Solidarity Association) was the first association based at the University of Ankara and aimed at protecting the rights of transgender individuals, continuing projects focused on discrimination, hate speech, violence, and social exclusion at both the national and international levels:

> *Apart from these organizations, LGBT activists can still have a role in the university campuses regarding the creation of student associations, organizing demonstrations with their schoolmates and colleagues in the university. Whether they belong to the LGBT community or not, they can share their experiences more closely without the fear of being judged and targeted by the regime.*[20]

Starting in 2006, several "Campus Meetings against Homophobia and Transphobia" were held within the three largest universities in Ankara and then extended to other universities. As one activist explains,

> *These were important occasions to raise awareness of LGBT people's struggles for rights, facing, for example, the issue of the medicalization of sexuality and of the conservative policies that justify the institutionalization of discrimination and inequality under the guise of terms like 'family values,' 'obscenity,' and 'public morals.'*[21]

The topics debated in these meetings included a wide range of issues:

> *We discussed every kind of topic, from the right to housing to gendered public space; from the right to work to the union movements to Social Service Areas for LGBT; from the discrimination of sexual orientation and identity in education to the right to public health and the trajectories of homosexuality and the LGBT Movement Against Inequality.*[22]

After the Gezi protests, Turkish LGBT activists increasingly relied on university campuses to organize and mobilize. Mobilization processes have been characterized by the presence of both structured organizations that have undertaken advocacy work for years and grassroots groups in which LGBT individuals have found space to participate and contribute to awareness-raising strategies.

Protests and LGBT activism in Egypt, Tunisia, and Turkey (2011–2013)

After 2013, Egypt and Tunisia showed divergent profiles. In Tunisia, a democratic transition culminated in the adoption of a new Constitution on January 14, 2014. Despite relevant achievements for Tunisian women, such as the approval of several laws increasing women's political and legal rights, including Law 58 criminalizing violence against women in 2017, since 2013 the Tunisian LGBT community has had to cope with widespread repressive campaigns and legislation. Article 230 of the Tunisian Penal Code, approved in 1913, provides up to three years for private acts of "sodomy" between consenting adults. After the 2010–2011 uprisings, the Tunisian Supreme Court stopped an attempt to cancel Article 230 of the Penal Code that criminalizes homosexuality. As one LGBT activist explains,

> Both the Islamists of Ennahda and the 2019 elected president, Kais Saied, strongly opposed the reform.[23]

Another LGBT activist who took part in protests in Tunis added,

> During major demonstrations, we always face threats and attacks by police officers. I personally witnessed several episodes of arbitrary arrests, harassments, and criminalization perpetrated by Tunisian policemen toward LGBT activists.[24]

Despite this harsh situation, the last few years have witnessed some mild openings in the Tunisian political and discursive opportunity structures (Antonakis 2019), such as the demands of banning forced anal tests as a violation of human rights in 2017[25] and the presence of the first openly homosexual presidential candidate, Shams president Munir Baatur, in the 2019 elections.[26] In April 2020, Baatur announced the alleged first gay marriage in the country. However, these openings are not without contradictions, since Baatur has been strongly criticized by many Tunisian activists:

> He used LGBT campaigns as a tool to increase his personal visibility; he is not a leader for the whole community.[27]

Furthermore, the Minister of Local Affairs, Lotfi Zitoun, publicly denied that Tunisian authorities legally recognized same-sex unions. In Egypt, state homo-transphobia had targeted LGBT individuals before the 2011 protests (Habib 2019; Ammar 2011). One of the most well-known repressive roundups took place on May 11, 2001, when police and state security officers raided the Queen Boat, anchored on the Nile, and arrested over 50 people on charges of "male prostitution." According to human rights activists, the detainees were physically and psychologically humiliated:

In those years, the Muslim Brotherhood was accusing the Mubarak regime of incompetence against anti-Islamic tendencies within the society. Thus, the LGBT community has been the first target of the authorities to silence the Muslim Brotherhood's accusations.[28]

Another activist highlighted that:

LGBT rights were not tackled during the Morsi presidency [the first elected Muslim Brotherhood president] between 2012 and 2013. I personally didn't take part in the electoral process, but I supported the Revolutionary Socialists.[29]

After the 2013 military coup and with the control over state institutions by incumbent president Abdel Fattah al-Sisi, the regime has strongly repressed political oppositionists, human rights advocates, and the Egyptian LGBT community, as an Egyptian LGBT activist explained:

Egyptian authorities are trying to demonstrate their opposition to any kind of anti-Islamic behaviors present in the country in order to lever on this and keep repressing Islamist movements.[30]

The places mentioned for gatherings include a few ancient hammams still located in downtown Cairo, from the environs of the market of the Bab el-Louk neighborhood to the old Bab Shareya Turkish bath and the cinemas in the poor neighborhood of Boulaq Abul-Ela. In 2014, the Egyptian police raided the Turkish bath "Sea Door" in the Ramsis neighborhood of Cairo; 33 people were subsequently accused of "immorality" and detained. Their arrests were covered extensively by Egyptian state television.

According to our interviewees, after the military coup, hundreds of LGBT activists have been imprisoned, chased away from their homes, or have lost their jobs due to their sexual orientation.[31] In November 2014, eight men were condemned to three years of detention on the charge of "debauchery":

They arrested them because they appeared in a video that, according to the authorities, represented a homosexual marriage on a boat on the Nile, but there was no evidence that the video was about a wedding.[32]

According to our interviewees, the men in this case were also subjected to anal tests. In the summer of 2014, police raided a house where a group of transgender individuals lived; in 2017, 16 LGBT activists were arrested in Egypt for waving a rainbow flag during a concert of the Lebanese band Masrou Leila in Cairo.

The arrested activists were charged with "inciting debauchery" and "abnormal sexual relations," and they were tortured in prison. The LGBT Egyptian activist Sarah Hegazy, who was among them, committed suicide in Canada three years later, where she had moved after being released. Her death strongly affected the Egyptian LGBT community:

> *I have been hiding all my life. When I knew about her death, I thought there was no reason to continue fighting.*[33]

In contrast to the LGBT mobilizations in Egypt and Tunisia, the Turkish LGBT movement dates back to the 1980s, paving the way for the flourishing of LGBT-based protests in the 1990s and 2000s (Ceylan 2015). In the last two decades, such mobilizations have become increasingly visible. Some associations and national NGOs have begun to work on the issue of discrimination and the exclusion of sexual minorities from society, despite the obstacles advanced by the Ministry of the Interior to block their legal recognition.

Although it is clear that the situation has worsened in recent years for those who support equality in Turkey (Göçmen & Yilmaz 2017), numerous organizations still focus on the protection and enhancement of sexual minorities' rights. One of them is Kaos-Gl, founded in Ankara in 1994 and registered in 2005. Another important Turkish LGBT association is Lambda, founded in 1993 in Istanbul. Lambda was among the organizers of the 2003 Pride, the first LGBT march in Turkey.

In 2008, a court decision banned the organization, assuming that its activities were "against laws and morality," but the Supreme Court overruled the decision. The LGBT activists we interviewed mentioned several episodes of repression and violence against the LGBT community.[34]

An example is the case of Ahmet Yildiz, a student at the University of Marmara who was killed in 2008 by his family because of his homosexuality. As a result of the Gezi protests,

> *the government made gatherings and having public assemblies for the LGBT community in Turkey even harder.*[35]

For instance, in 2015, Turkish authorities assigned police to contain the protests, which used tear gas and water cannons to disperse the Istanbul Gay Pride march, claiming that it had been organized during a day of Ramadan.[36]

Thus, these three countries show that, despite differences in the sociopolitical and cultural contexts in which LGBT activists are embedded, the rise to power of repressive and conservative regimes not only challenges sexual minorities' mobilizations and visibility in the public sphere but also threatens their very existence and survival.

LGBT activism in repressive contexts: strategies to resist and exist

Participation in the 2011 and 2013 protests has only slightly improved LGBT activists' ability to make their claims heard in the public sphere. In response, they have often been targeted by public authorities as a threat to traditional values, pushing LGBT communities not only to mobilize but also to articulate innovative strategies to resist and exist under hostile conditions.

In Egypt, both state repression and stigmatization remain forceful in the public and private local media. For example, Patrick Zaki, a student of Gender Studies at the University of Bologna who was arrested on his arrival in Cairo on February 7, 2020, and is still in prison on charges of acting against national interests through cyber-activism, has been stigmatized by media commentators because he went abroad "with the aim to study homosexuality." As one activist highlighted,

> *The Egyptian authorities still portray LGBT activism as a 'foreign form of activism,' exported to Egypt.*[37]

In the last few years, the Egyptian police have expelled several homosexual foreigners, tourists, and resident individuals, preventing them from returning to the country. However, queer life in Cairo and Alexandria is by no means impossible, especially thanks to dedicated smartphone applications (Alqaisiya 2020):

> *Despite repression and control over cyber-activism, I can still have my queer life in Cairo, meeting people and building relationships, thanks to Grindr [a gay chat application] and other apps.*[38]

Hence, new technologies have played a role in the repertoires of action articulated by LGBT Egyptian activists: on the one hand, communication platforms are deployed to organize and coordinate informal gatherings; on the other hand, social media and apps provide individuals with the possibility to meet other LGBT people, thus strengthening processes of community building.

As in Egypt, Tunisian LGBT activists have had to face increased state homo-transphobia after the protests:

> *This recently achieved visibility for LGBT activists, and the cooperation of these social movements triggered the anger of the police. LGBT activists are now a target for constant police intimidation; their pictures are often published on social networks, together with insults and death threats. An example in this instance is the LGBT activist Rania Amdouni, who has been harshly intimidated by the police syndicates.*[39]

However, Tunisian LGBT individuals still meet at gay bars, cafes, and dance halls, which are among the most fashionable places for leisure in Tunis:

> *Wax, Yuka, and Habibi are among the best LGBT friendly bars and discos in the Gamra neighborhood in Tunis, opened in recent years. These*

> *new bars are places for gatherings for Tunisian gays and lesbians, but they are spaces of freedom for the whole Tunisian youth,*[40]

as an LGBT activist who took part in protests in Tunis explained. As the Tunisian case shows, under repressive regimes, not only social media, but also venues such as cafes, restaurants, and bars may represent spaces where marginalized communities have the possibility to share their experiences and create a community that provides the base to mobilize for their rights through more structured advocacy organizations.

In Turkey, after the failed coup attempt on July 15, 2016, President Erdoğan and the Turkish authorities have been waging a war against alleged oppositionists (e.g., policemen, judges, journalists, civil servants), stigmatizing LGBT activists as 'deviants.'[41] As an LGBT activist explains,

> *Their political discourse uses Islam in order to claim that things are 'not Islamic enough,' and universities, as possible 'spaces for opposition,' are among the first to be censored.*[42]

And another activist highlights that

> *LGBT people witness employment refusals or are even fired on the basis of their sexual orientation. Discriminatory laws make it impossible for them to practice their professions or even look for fair trials after that.*[43]

This is echoed in the words of another activist:

> *Although in the last two decades, the LGBT movements in Turkey have moved forward in terms of organization and visibility in society, they continue to be disrupted by the authoritarianism of the regime.*[44]

Against this backdrop, in recent years, campuses have functioned as 'safer' spaces[45] for LGBT people persecuted by the regime:

> *Recently, things have been severely deteriorating: the reality of LGBT individuals is of a constant deprivation of fundamental human rights, and this is happening very fast,"*[46] *or "Authorities in Ankara are imposing a ban on all LGBT cultural events, citing threats to public order and fear of 'provoking reactions within certain segments of society."*[47]

Moved by a *"great desire to make LGBT solidarity visible,"*[48] university campuses and organizations have thus been providing spaces both to make LGBT claims visible through public meetings and discussions and to create a stronger and more resilient LGBT community.

Conclusion

This chapter examined the evolution of the repertoires of action articulated by LGBT activists based in Egypt, Tunisia, and Turkey since the 2011 and 2013 protests. Empirically, the study drew on a qualitative analysis using data collected through semi-structured interviews undertaken between 2011 and 2020 in Egypt, Tunisia, and Turkey. All three countries have indeed witnessed a constant presence of LGBT activists during recent anti-regime demonstrations that broke out in Egypt before and after 2011, in Tunisia before and in the aftermath of the 2010–2011 uprisings, and in Turkey before and in the aftermath of the Gezi Park movement (2013).

Our findings showed that LGBT communities have been constantly and disproportionately targeted by state and non-state repressive campaigns (e.g., police, army, Islamists) in all three countries. However, they have articulated different repertoires of action. In Egypt, LGBT activists have challenged repression by using social networks as alternative venues for socialization due to the lack of organized LGBT groups. On the contrary, the presence of stronger meso-level mobilizing structures (e.g., student organizations within university campuses, or post-uprising legalized LGBT associations) have helped Tunisian and Turkish activists to take advantage of their new visibility to increase cooperation with other political challengers within the framework of new waves of protests. Furthermore, we found that CSOs use social media as a means to navigate their struggle for (in) visibility. On the one hand, these platforms serve to inform about the aims of LGBT organizations, as well as to publicly defy homophobic behaviors committed by state and non-state actors. On the other hand, ICTs also constitute private networks for LGBT individuals to build local communities. Social media were deployed for both moderate actions, such as visible advocacy, and for more depoliticized actions, for instance, as venues for safer encounters to resist increasing repression.

Future research may strengthen the comparative dimension of these results by examining both the evolution of LGBT activism in the three examined countries and LGBT individuals' participation in protests, repertoires of action, alternative use of social networks, and meso-level organizations in other neighboring countries (e.g., Algeria).

Notes

1 This chapter has been previously published in a different format in Social Movement Studies Acconcia, G., Perego, A., & Perini, L. (2022). LGBTQ activism in repressive contexts: The struggle for (in)visibility in Egypt, Tunisia, and Turkey. *Social Movement Studies*. doi: 10.1080/14742837.2022.2070739.

2 Lesbian, gay, bisexual, trans, queer, intersex, asexual, and other (LGBTQIA*) issues and identities have been addressed by scholars in diverse ways. For the purpose of this chapter, we mainly use the umbrella terms 'LGBT' and 'sexual minorities' to refer to people marginalized because of sexual orientations and/

or gender identities that are deviant from cis-heteronormative frameworks. The acronym 'LGBT' was also used by most of the interviewees to refer to their own identities and activism.

3 The few investigations on LBGT activism in MENA during and after the 2011 uprisings may be complemented by studies on how mobilizations against gender-based violence and sexual harassment have evolved in the region since the Arab Spring (Chafai 2020; El Ashmawy 2017; Rizzo et al. 2012).

4 A third hypothesis concerns the diffusion of protests beyond national borders (Tarrow 1996). This hypothesis will nonetheless not be considered in our paper, since our research focus concerns the innovation of LGBT repertoires of action within the national context.

5 The term 'civil society organizations' includes both formal organizations and informal groups (Edwards 2004). Regardless of their level of formalization, both are civil society actors "promoting collective action on public issues, whether on a service delivery or a protest-oriented basis" (Diani 2015: 35).

6 In contrast to scholars who emphasize the benefits brought by ICTs to collective efforts, some researchers have shown that social media have also been exploited by state authorities to identify and monitor activists in repressive contexts (Michaelsen 2017; Xu 2021). More on digital surveillance in authoritarian contexts at: www.jadaliyya.com/Details/34672 [Last accessed September 20, 2021].

7 Drawing on Habermas (1989), the concept of the public sphere is here to be understood as a discursive arena where different publics engage in discussions and contestations.

8 More information available at: http://turkishpolicy.com/article/879/turkeys-for-ays-into-the-middle-east [Last accessed May 24, 2021].

9 In Egypt and Tunisia as well, university campuses have been used as venues for youth movements. For example, supporters of the Revolutionary Socialists, including individual LGBT activists, formed the National Alliance for Change and Unions within universities in 2005 (Acconcia & Pilati 2021).

10 Interviewee 20, Cairo, 2015.

11 Interviewee 7, Cairo, 2013.

12 Interviewee 2, Cairo, 2012.

13 Interviewee 1, Cairo, 2011.

14 Interviewee 14, Cairo, 2014.

15 Interviewee 28, Tunis, 2019.

16 Interviewee 30, Tunis, 2019.

17 Interviewee 36, Tunis, 2019.

18 The Gezi Park movement started as a protest against government plans to rebuild Ottoman barracks and a shopping mall on the edge of Taksim Square. This decision entailed a dramatic escalation of events including the stigmatization of protesters as terrorists, arrests, and enforced exiles.

19 Interviewee 42, Istanbul, 2017.

20 Interviewee 41, Istanbul, 2017.

21 Interviewee 42, Istanbul, 2017.

22 Interviewee 44, Istanbul, 2017.

23 Interviewee 31, Tunis, 2019. According to the incumbent Tunisian president, the LGBT community is "receiving funds from abroad to corrupt the Islamic nation."

24 Interviewee 36, Tunis, 2019. See also www.hrw.org/news/2021/02/23/tunisia-police-arrest-use-violence-against-lgbti-activists [Last accessed September 21, 2021].

25 www.hrw.org/news/2017/05/03/consent-or-no-anal-testing-tunisia-must-go

26 The electoral committee rejected his candidacy without providing details.

27 Interviewee 34, Tunis, 2019.
28 Interviewee 6, Alexandria, 2015.
29 Interviewee 3, Cairo, 2012.
30 Interviewee 15, Cairo, 2014.
31 See also www.hrw.org/news/2020/10/01/egypt-security-forces-abuse-torture-lgbt-people [Last accessed September 21, 2021].
32 Interviewee 27, Cairo, 2015.
33 Interviewee 28, Alexandria, 2020.
34 Interviewees 41 and 42, Istanbul, 2017.
35 Interviewee 40, Istanbul, 2017.
36 Information available at: www.reuters.com/article/us-turkey-rights-pride-idUSKCN0P8OOQ20150628 [Last accessed May 3, 2021].
37 Interviewee 24, Cairo, 2019. See also www.hrw.org/news/2020/03/20/egypts-denial-sexual-orientation-and-gender-identity [Last accessed September 21, 2021].
38 Interviewee 23, Cairo, 2019.
39 Interviewee 32, Tunis, 2019.
40 Interviewee 29, Tunis, 2019.
41 www.bbc.com/news/world-europe-55901951
42 Interviewee 41, Istanbul, 2017.
43 Interviewee 40, Istanbul, 2017. See also www.hrw.org/news/2021/03/24/turkey-erdogans-onslaught-rights-and-democracy [Last accessed September 21, 2021].
44 Interviewee 39, Istanbul, 2017.
45 The concept of safe space emerged in both feminist and LGBT groups of the 1960s and 1970s (Kenney 2001) and has since been developed by both activists and scholars (The Roestone Collective 2014). Safe spaces can be understood as venues – either physical, digital, or symbolic – where marginalized individuals can feel free from violence and harassment (The Roestone Collective 2014). Within this framework, safe spaces are thought to enhance collective strength, thus generating strategies for resistance to political or social repression (Kenney 2001: 24). The concept has nonetheless been highly debated. Black and intersectional feminists have pointed out how so-called 'safe spaces' have often reproduced unequal power relations that particularly affect individuals who are marginalized along various inequality lines, such as black women or LGBT migrants. Building on these insights, activists have hence started to use the expression 'safer spaces' to acknowledge that such venues are not immune to power relations. See also https://splinternews.com/what-s-a-safe-space-a-look-at-the-phrases-50-year-hi-1793852786 [Last accessed September 10, 2021].
46 Interviewee 43, Istanbul, 2017.
47 Interviewee 42, Istanbul, 2017.
48 Interviewee 39, Istanbul, 2017.

References

Acconcia, G. (2018). *The uprisings in Egypt: Popular committees and independent trade unions* (PhD thesis). London: Goldsmiths College, University of London.
Acconcia, G., & Pilati, K. (2021). Variety of groups and protests in repressive contexts: The 2011 Egyptian uprisings and their aftermath. *International Sociology, 36*(1), 91–110.

Almeida, P. D. (2003). Opportunity organizations and threat-induced contention: Protest waves in authoritarian settings. *American Journal of Sociology, 109*(2), 345–400.

Alqaisiya, W. (2020). Guest editor's introduction: Queerness with Middle East Studies: Mapping out the useful intersections. *Middle East Critique, 29*(1), 3–7.

Ammar, P. (2011). Middle East masculinity studies: Discourses of 'Men in Crisis,' industries of gender in revolution. *Journal of Middle East Women's Studies, 7*(3), 37–70.

Antonakis, A. (2019). *Renegotiating gender and the state in Tunisia between 2011 and 2014*. Berlin: Springer.

Ayoub, P. M. (2016). *When states come out. Europe's sexual minorities and the politics of visibility*. Cambridge: Cambridge University Press.

Bayat, A. (2002). Activism and social development in the Middle East. *International Journal of Middle East Studies, 34*, 1–28.

Beinin, J., & Vairel, F. (2011). *Social movements, mobilization, and contestation in the Middle East and North Africa*. Stanford, CA: Stanford University Press.

Bernstein, M. (1997). Celebration and suppression: The strategic uses of identity by the lesbian and gay movement. *American Journal of Sociology, 103*(3), 531–65.

Birdal, M. S. (2020). The state of being LGBT in the age of reaction: Post-2011 visibility and repression in the Middle East and North Africa. In M. J. Bosia, S. M. McEvoy, & M. Rahman (Eds.), *The Oxford handbook of global LGBT and sexual diversity politics* (pp. 267–280). New York, NY: Oxford University Press.

Buyantueva, R., & Shevtsova, M. (2019). Introduction: LGBTQ+ activism and the power of locals. In R. Buyantueva & M. Shevtsova (Eds.), *LGBTQ+ activism in central and Eastern Europe: Resistance, representation, and identity* (pp. 1–19). London: Palgrave Macmillan.

Ceylan, E. (2015). LGBT in Turkey: Policies and experiences. *Social Sciences, 4*(3), 838–858.

Chafai, H. (2020). Everyday gendered violence: Women's experiences of and discourses on street sexual harassment in Morocco. *The Journal of North African Studies, 26*(5), 1013–1032.

Clark, J. A. (2004). *Islam, charity, and activism: Middle-class networks and social welfare in Egypt, Jordan, and Yemen*. Bloomington, IN: Indiana University Press.

Currier, A. (2012). *Out in Africa: LGBT organizing in Namibia and South Africa*. Minneapolis, MN: University of Minnesota Press.

Davenport, C. (2005). Repression and mobilization: Insights from Political Science and Sociology. In C. Davenport, H. Johnston, & C. Mueller (Eds.), *Repression and mobilization* (pp. vii–xli). Minneapolis, MN: University of Minnesota Press.

Davenport, C. (2007). State repression and political order. *Annual Review of Political Science, 10*, 1–23.

D'Emilio, J. (1983). *Sexual politics, sexual communities*. Champaign, IL: University of Chicago Press.

Diani, M. (2015). *The cement of civil society. Studying networks in localities*. New York, NY: Cambridge University Press.

Dorsey, J. M. (2012). Pitched battles: The role of ultra soccer fans in the Arab spring. *Mobilization, 17*(4), 411–418.

Duboc, M. (2011). Egyptian leftist intellectuals' activism from the margins: Overcoming the mobilization/demobilization dichotomy. In J. Beinin & F. Vairel (Eds.),

Social movements, mobilization, and contestation in the Middle East and North Africa (pp. 61–80). Stanford, CA: Stanford University Press.

Earl, J. (2003). Tanks, tear gas, and taxes: Toward a theory of movement repression. *Sociological Theory, 21*, 44–68.

Edenborg, E. (2017). *Politics of belonging.* New York, NY: Routledge.

Edenborg, E. (2020). Visibility in global queer studies. In M. J. Bosia, S. M. McEvoy, & M. Rahman (Eds.), *The Oxford handbook of global LGBT and sexual diversity politics* (pp. 349–366). New York, NY: Oxford University Press.

Edwards, M. (2004). *Civil society.* Cambridge: Polity Press.

El Amrani, F. Z. (2019). Minority-ness in the Post-Arab spring discourse: LGBT community in the 20th February movement. In H. Tayebi & J. Lobah (Eds.), *Dynamics of inclusion and exclusion in the MENA region: Minorities, subalternity, and resistance* (pp. 273–290). Rabat: Hanns Seidel Foundation Morocco/ Mauritania.

El Ashmawy, N. (2017). Sexual harassment in Egypt class struggle, state oppression, and women's empowerment. *Journal of Women of the Middle East and the Islamic World, 15*, 225–256.

El Menyawi, H. (2006). Activism from the closet: Gay rights strategizing in Egypt. *Melbourne Journal of International Law, 7*(1), 28–51.

Ferree, M. M. (2004). Soft repression: Ridicule, stigma, and silencing in gender-based movements. In D. Meyers & D. Cress (Eds.), *Authority in contention (research in social movements, conflicts and change, Volume 25)* (pp. 85–101). London: Emerald Group Publishing Limited.

Fortier, E. A. (2015). Transition and marginalization: Locating spaces for discursive contestation in post-revolution Tunisia. *Mediterranean Politics, 20*(2), 142–160.

Fortier, E. A. (2019). *Contested politics in Tunisia: Civil society in a post-authoritarian state.* Cambridge: Cambridge University Press.

Göçmen, I., & Yilmaz, V. (2017). Exploring perceived discrimination among LGBT individuals in Turkey in education, employment, and health care: Results of an online survey. *Journal of Homosexuality, 64*(8), 1052–1068.

Goldstone, J., & Tilly, C. (2001). Threat (and opportunity): Popular action and state response in the dynamic of contentious action. In R. Aminzade, J. Goldstone, D. McAdam, E. Perry, W. Sewell, & S. Tarrow (Eds.), *Silence and voice in the study of contentious politics* (pp. 179–194). Cambridge: Cambridge University Press.

Gorkemli, S. (2012). "Coming out of the internet": Lesbian and gay activism and in the internet as a "Digital closet" in Turkey. *Journal of Middle East Women's Studies, 8*(3), 63–88.

Habermas, J. (1989). *The structural transformation of the public sphere: An inquiry into a category of Bourgeois society.* Cambridge: Polity Press.

Habib, S. (2019). Queer names and identity politics in the Arab world. In *The global encyclopedia of lesbian, gay, bisexual, transgender, and queer history.* New York, NY: Scribner.

Hamanaka, S. (2020). The role of digital media in the 2011 Egyptian revolution. *Democratization, 27*(5), 777–796.

Hassenpour, N. (2014). Media disruption and revolutionary unrest: Evidence from Mubarak's quasi-experiment. *Political Communication, 31*(1), 1–24.

Howard, P. N., & Hussain, M. M. (2011). The role of digital media. *Journal of Democracy, 22*(3), 35–48.

Howard, P. N., & Hussain, M. M. (2013). *Democracy's fourth wave? Digital media and the Arab Spring*. Oxford: Oxford University Press.

Jamal, A. (2007). *Barriers to democracy: The other side of social capital in Palestine and the Arab world*. Princeton, NJ: Princeton University Press.

Kenney, M. R. (2001). *Mapping gay L.A.: The intersection of place and politics*. Philadelphia, PA: Temple University Press.

Magued, S. (2021). The Egyptian LGBT's transnational cyber-advocacy in a restrictive context. *Mediterranean Politics*. doi: 10.1080/13629395.2021.1905924

Massad, J. (2002). Re-orienting desire: The gay international and the Arab world. *Public Culture, 14*(2), 361–385.

Michaelsen, M. (2017). Far away, so close: Transnational activism, digital surveillance and authoritarian control in Iran. *Surveillance and Society, 15*(3/4), 465–470.

Needham, J. (2013). After the Arab spring: A new opportunity for LGBT human rights advocacy? *Dule Journal of Gender, Law, and Policy, 20*, 287–323.

O'Dwyer, C. (2018). *Coming out of communism: The emergence of LGBT activism in Eastern Europe*. New York, NY: New York University.

Pilati, K. (2016). Do organizational structures matter for protests in nondemocratic African countries? In E. Y. Alimi, A. Sela, & M. Snajder (Eds.), *Popular contention, regime, and transition: The Arab revolts in comparative global perspective* (pp. 46–72). New York, NY: Oxford University Press.

Rasler, K. (2016). Understanding dynamics, endogeneity, and complexity in protest campaigns: A comparative analysis of Egypt (2011) and Iran (1977–1979). In E. Y. Alimi, A. Sela, & M. Snajder (Eds.), *Popular contention, regime, and transition: The Arab revolts in comparative global perspective* (pp. 180–202). New York, NY: Oxford University Press.

Rizzo, H., Price, A. M., & Meyer, K. (2012). Anti-sexual harassment campaign in Egypt. *Mobilization, 17*(4), 457–475.

Sherif, S. (2020). Transgender visibility/invisibility: Navigating cisnormative structures and discourses. *Kohl Journal, 6*(3). Retrieved September 20, 2021, from https://kohljournal.press/transgender-visibility-invisibility

Tarrow, S. (1996). States and opportunities: The political structuring of social movements. In D. McAdam, J. McCarthy, & M. Zald (Eds.), *Comparative perspectives on social movements* (pp. 41–61). Cambridge: Cambridge University Press.

Tétreault, M. A. (2000). *Stories of democracy: Politics and society in contemporary Kuwait*. New York, NY: Columbia University Press.

The Roestone Collective. (2014). Safe space: Towards a reconceptualization. *Antipode, 46*(5), 1346–1365.

Tilly, C. (1986). *The contentious French*. Cambridge, MA: Harvard University Press.

Tilly, C. (2008). *Contentious performances*. Cambridge: Cambridge University Press.

Tschantret, J. (2020). Revolutionary homophobia: Explaining state repression against sexual minorities. *British Journal of Political Science, 50*, 1459–1480.

Van de Bildt, J. (2015). The quest for legitimacy in postrevolutionary Egypt: Propaganda and controlling narratives. *The Journal of the Middle East and Africa, 6*, 253–274.

Waaldjik, K. (1994). Standard sequences in the legal recognition of homosexuality – Europe's past, present and future. *Australasian Gay and Lesbian Law Journal, 4*, 50–72.

Wilkinson, C. (2014). LGBT human rights versus 'Traditional Values': The rise and contestation of anti-homopropaganda laws in Russia. *Journal of Human Rights, 13*(3), 363–379.

Wilkinson, C. (2017). Introduction: Queer/ing in/security. *Critical Studies on Security, 5*(1), 106–108.

Wilkinson, C. (2020). LGBT rights in the former Soviet Union: The evolution of hypervisibility. In M. J. Bosia, S. M. McEvoy, & M. Rahman (Eds.), *The Oxford handbook of global LGBT and sexual diversity politics* (pp. 233–248). New York, NY: Oxford University Press.

Xu, X. (2021). To repress or to co-opt? Authoritarian control in the age of digital surveillance. *American Journal of Political Science, 65*(2), 309–325

Yeşilada, B. A. (2015). The future of Erdoğan and the AKP. *Turkish Studies, 17*(1), 19–30.

Zivi, K. (2012). *Making rights claims: A practice of democratic citizenship.* New York, NY: Oxford University Press.

6 The Kurds of Syria

From Popular Committees to fighting units

By adopting SMT as a basic framework to analyze the 2011 uprisings in the MENA region, this chapter will examine the role of alternative networks and other forms of political conflict in reference to the Syrian Kurdistan case. The initial demonstrations in Northern Syria between 2011 and 2012 sparked the formation of new means of popular mobilization, and triggered mass participation in alternative networks aimed at recruiting ordinary citizens to provide social services, security, and self-defense. Drawing on interviews with participants in the People's Protection Units and Women's Protection Units (YPG-YPJ [*Yekîneyên Parastina Gel-Yekîneyên Parastina Jin*]) carried out during the author's stay in Syria in 2015, insights into the workings and attempts of institutionalization of the Popular Committees (*Mala Gel*) and Women Committees (*Mala Jin*) will be provided. In this chapter, it will be argued that in the context of war in Northern Syria between 2013 and 2016, with the emergence of a very diverse range of jihadist groups, including ISIS, participants within these Popular Committees felt the need to be involved in direct action, including the armed struggle, in order to protect their neighborhoods and replace the constant absence of security personnel. Thus, in Syria, these social movements evolved into paramilitary organizations that are very different compared to other grassroots mobilizations in the region.

Introduction

This chapter will focus on the workings and attempts of institutionalization of grassroots mobilizations in Northern Syria in reference to the emergence and evolution of the Popular Committees and Women Committees (*Mala Gel* and *Mala Jin*; Boothroyd 2016) in the canton of Kobane. The research draws on interviews with Syrian fighters carried out in this urban district during the author's stay in Syria in the spring of 2015. The levels of initial political participation of the components of the Popular Committees in Syria, the way in which they perceived insecurity and the need for stability, the reasons that triggered their mobilization and demobilization, the attempts to institutionalize the precarious structures of the Committees, and

DOI: 10.4324/9781003293354-7

the parallel evolution into armed paramilitary groups in the context of war will be addressed.

As Gelvin (1998) highlights, the formation of such grassroots mass movements in Syria emerged in Syria between 1918 and 1920. Local Committees were already a modern pattern of political mobilization and an example of attempts at a process of institutionalization by non-state elites (Gelvin 1998). King Faisal governed Syria for two years after World War I, between the end of the Ottoman Empire and the beginning of the French mandate (Gelvin 1994). In a more politicized public space, an autonomous civil society pushing for the accountability of state institutions was emerging. In this context of economic, administrative, and urban transformation, new relations of power were about to surface; as the domain of formal politics expanded, the state and the market provided new functions. At that stage, the Popular Committees began to be involved in a diverse number of modern and alternative activities, partially alien to the state apparatus. Local representatives were chosen through district elections (Gelvin 1994).

Moreover, the Popular Committees recruited their own militias. In every urban neighborhood, they began to hold meetings to coordinate the various militias' branches, which filled the vacuum left by the lack of governmental control.

A century ago, the Popular Committees in Syria were proposing new power relationships: first, they affirmed the prerogatives of their elected members through open elections; second, they demanded the accountability of the urban elites controlling their resources and functions; and third, they wanted to institutionalize the electoral procedures on a national base, forming a "supervisory committee" to oversee the citizens of every neighborhood. During this period, and as happened in some parts of the country between 2011 and 2013, an increasing number of ordinary citizens began to contest a growing number of public issues (Gelvin 1994).

In this regard, Gelvin (1998) argues that the Popular Committees "institutionalized and broadened horizontal and associational ties"; they defined a new framework for social and political legitimacy, filling "the void that neither the government" nor the "national organizations were structurally or ideologically capable of filling" (Gelvin 1998).

Popular Committees are "self-defense groups heterogeneous in their tactics, organization, and efficacy, but a critical response to the security vacuum" (Hassan 2015). This chapter argues that the Northern Syrian Popular Committees challenged traditional patterns of state control over civil society, building up a new political and paramilitary structure of power. I decided to focus on the Northern Syria Popular Committees, rather than Tunisia, Algeria, or Palestine, where the relevance of trade unions is historically central, because in this area the grassroots mobilizations, organized in local committees, appeared to have relevant parallels with the Egyptian case study from an early stage. However, as Ismail explains, subaltern actors in Egypt and Syria shaped their evolution differently in relation to their attitude toward the state. Compared to Egypt, in Syria, the regime's strategies of control, or *divide and rule*, over grassroots mobilizations manipulated subaltern forces. In other words, this process contributed to fragmenting the oppositionists, turning groups against each other.

However, it is not the aim of this chapter to discuss the dynamics of the Syrian civil war, but to disentangle the evolution of Popular Committees from a comparative perspective. In 2011, as el-Meehy states, the Popular Committees in Syria can be defined as "horizontal forms of committee-centered grassroots activism" (El-Meehy 2017), also known as *tanseeayat*, or "ad hoc local coordination committees" (El-Meehy 2017). Among the activities of the Syrian local committees in the initial stages of mobilization, "they extended support for families of prisoners, provided emergency relief to internally displaced persons, and committed local armed groups to sign up to an ethical code of conduct for observing human rights" (El-Meehy 2017).

According to el-Meehy, the first local committees were founded in Aleppo and Al-Zabadani. Although they quickly spread across the country, "by 2016, the number of active councils had fallen sharply to around 395" (El-Meehy 2017), with the majority located in Northern Syria. The initial demonstrations and riots in Northern Syria between 2011 and 2012 sparked the formation of new means of popular mobilization, and triggered mass participation in alternative networks that aimed to recruit ordinary citizens to provide social services, security, and self-defense.

Later, in the context of the war in Northern Syria between 2013 and 2016, with the further emergence of a very diverse range of jihadist groups (including ISIS), the participants within those Popular Committees felt the need to be involved in direct action, including the armed struggle, in order to protect their neighborhoods and substitute for the constant absence of security personnel, thereby defending their properties from the attacks from a wide range of both oppositional groups and regime supporters.

Thus, in the Syrian War, these social movements evolved into paramilitary organizations that were very different compared to the Egyptian case study where, in a context of diminishing mobilization, they became closer to private voluntary organizations than to revolutionary groups, as previously argued. At that stage, in Northern Syria, the Committees were pivotal in forming armed entities, such as the People's Protection Units and Women's Protection Units, which began to provide systems of patrols to guarantee local and external security, building up an embryonic autonomous government.

Kobane Canton

This research focuses on the urban district of Kobane and its surroundings due to the sufficient level of security present after ISIS withdrew in January 2015, the proximity to the Turkish border of this relatively accessible region compared to other parts of Syria, and the high levels of politicization and mobilization among the Syrian Kurds.

With the end of French colonization, the Syrian Kurds in the three provinces of Jazira, Efrin, and Kobane were both excluded from Northern Turkish Kurdistan and isolated by their neighbors' growing Arab nationalism (Allsopp 2015). Later, as a consequence of the Hasaka Census (1962), thousands of Syrian Kurds were left without citizenship and excluded from the labor market. At that stage, the major local pro-Kurdish left-communalist parties were founded along with 32 other smaller leftist political groups.

Those activists mobilized very little support until the aid given by the United States to the Iraqi Kurdish fighters (*peshmerga*) between 2003 and 2005. With the 2012–2013 uprisings, the Syrian Kurds joined the opposition to the Assad regime within the framework of the Damascus Declaration issued in October 2005. However, with the 2011–2012 uprisings in Syria, the PYD joined neither the Kurdish coalition (KNC) nor the Arab opposition groups; instead, it began to implement Abdullah Öcalan's theories (Bookchin 2015) of democratic autonomy, forging self-defense groups and organizing an armed wing: the YPG/YPJ.

Compared to mid-20th century approaches to guerrilla warfare (e.g., Mao, Guevara), the Kurdish communalists provided a non-violent critique of hierarchical and capitalist societies. As Bookchin (2015) explains, to define his notion of "libertarian municipalism":

> *Communalism seeks to recapture the meaning of politics in its broadest, most emancipatory sense to fulfill the historic potential of the municipality as the developmental arena of mind and discourse. It conceptualizes the municipality, potentially at least as a transformative development beyond organic evolution into the domain of social evolution. The city is the domain where the archaic blood tie that was once limited to the unification of families and tribes, to the exclusion of outsiders, was – juridically, at least – dissolved. It became the domain where hierarchies based on parochial and sociobiological attributes of kinship, gender, and age could be eliminated and replaced by a free society based on a shared common humanity.*

In Northern Syria, popular assemblies have been organized; local councils have been formed in respect to ethnic and gender differences in cooperation with the Kurdistan Workers' Party (PKK).[1] The PYD has always fought in Syria and not in other adjacent countries, and its autonomous struggle has been characterized by supporting neither Assad nor the rebel opposition, instead taking a pragmatic and situational position depending on what would best benefit their cause (Allsopp & Acconcia 2013). On the one hand, moderate Arab oppositional groups have appeared to be hostile to the Syrian Kurds' demands. They have often accused the PYD of being in agreement with Assad against the Free Syrian Army (FSA). On the other hand, the PYD have accused all the anti-Assad militias of working in coordination with the Turkish army.

It is not the aim of this research to discuss geopolitical alliances in the region. However, it is relevant to add here that, despite the worsening crisis in 2014, with ISIS conquering the Northern Syrian region and Iraq, the Turkish government has been accused by the YPG/YPJ of delivering weapons and fighters to ISIS across the Syrian border, while never supporting the PYD due its link with the PKK; at the same time, the Syrian Kurdish fighters have been criticized by the Kurdistan Democratic Party (KDP) in Iraq and some human rights organizations (e.g., Amnesty International) as utopian and exclusionary (Graeber 2017). However, the YPG/YPJ gained some support from the US coalition against ISIS due to its effective combat performance (Acconcia 2015).

Still, few academic studies have focused on the development of the Popular Committees in the ongoing civil war in Northern Syria. In doing so, this chapter will try to add a better understanding of the development of grassroots' mobilizations in an urban district, considering them as long-term phenomena of political mobilization.

Methodology

In this research, 12 young Syrian Kurds who joined the YPG/YPJ and two unit commanders were interviewed. Diane and Rangin were the gatekeepers interviewed to select the YPG/YPJ fighters involved in this focus group. First, we met YPG Commander Diane at the Tall Abyad frontline in June 2015, a few days before the liberation of the town from ISIS supporters. Diane spent his youth in Lebanon in contact with Abdullah Öcalan when he fled to Syria (1980–1998). He had been a YPG commander during the 2014 battle for the liberation of Kobane. Later, we met YPJ Commander Rangin at YPJ's headquarters in Kobane, a few days after the liberation of Tall Abyad in June 2015. Rangin joined the YPJ in 2013; she was previously a journalist and part of the management of a local institution supporting women's rights.

In the preliminary stage, Diane and Rangin were part of the process of composing and organizing this specific focus group. The aim was to choose male and female members of YPG/YPJ (with or without previous involvement in the Popular Committees) in order to analyze the evolution of these local social movements before and after the ISIS occupation of Kobane (2014), the reasons and degrees of mobilization and demobilization within the social movements in the context of peace and cooperation with other oppositional groups, the levels of political participation, their own personal accounts of the ISIS invasion of Northern Syria in 2014, the relations between male and female members of the groups, and the evolution of civilian movements into a paramilitary and then a full-fledged armed force. At the end of the two meetings, we held a debriefing session with the gatekeepers to talk about the group dynamics and the results of these sessions.

Maslum, 22 years old, injured five times in the battles with ISIS, was unemployed in Lebanon, and joined YPG as a *heremi* (local self-defense citizen) in 2013, while his parents fled to Turkey. Agid, 18 years old, the eldest son of a farming family, worked as a carpenter in Istanbul before joining the YPG. Chakdar, 23, was from Kobane; his parents were farmers while he ran an electronics shop before joining the YPG. Chaidar, 21, from Tall Abyad, had joined the YPG two years previously. His parents were unemployed and living in Kobane. He worked in a mobile phone shop, and one of his brothers had been killed by ISIS. Biriar, 19, joined the PYD's youth group before entering the YPG. His mother was working in a center supporting the relatives of people killed by ISIS. Zenar, 19, was a tailor and had joined the YPG six months earlier. Massum, 29, a former construction worker, had joined the YPG two years before. Rashuan, a 31-year-old house painter, had joined the YPG one year previously. Filos, 29, from Kobane, had a brother who owned a minimarket; he was one of seven brothers and sisters, three of whom were working abroad. He was a welder before joining the YPG. Nurshin, 18, from Kobane, had been a student before joining the YPJ three months previously. Cicak, 25, was a student; her parents were farmers. Polda, 19, had been fighting with the YPG for one year.

The interviews were conducted in two long, distinct sessions at the Tall Abyad frontline and the YPJ's headquarters in Kobane in June 2015. From the very beginning, a promising level of interest was found among these male and female YPG/YPJ fighters. However, a number of problematic issues had to be tackled at a later stage: for example, in terms of access to the field during the periods of armed combat, timing related to the daily shifts of the soldiers, the presence of mines, and the aggressive methods of the Turkish authorities in refusing to grant official permissions to go to Syria. We met the YPG/YPJ fighters in their camp at the Tall Abyad frontline and the Kobane headquarters in June 2015, before and after the long struggle for the liberation of Tall Abyad from ISIS. While in the first stage, access to fighters, both male and female, was constantly mediated by the unit commanders who convinced them to be fully involved in the research, despite the ongoing daily fight against ISIS, the interviewees were later independently involved in the focus group.

At the beginning of this research, the interviewees expressed no security concerns regarding their participation in the research. Even at a later stage and after the liberation of Tall Abyad, the interviewees bore no personal concerns, fears, or mistrust. This research was informed by data collected in northern Syria (Rojava) in 2015. After conducting the fieldwork research, I was arrested at the border on my way back to Turkey after the interviews with YPG-YPJ fighters, together with other foreigners.[2] In the end, we were expelled by the Turkish authorities after two days of detention. It is evident that one of their priorities is to prevent any kind of coverage and research centered on what is happening in Rojava and the Kurdish areas of Turkey.

Hierarchy and organization

I will further describe the evolution of the Popular Committees in Northern Syria and the workings of a YPG/YPJ unit in Kobane, in reference to the interviews from our focus group. I will further highlight the extent to which, in the context of war, voluntary networks of self-defense groups, forged in an increasing environment of political participation, evolved into a more structured military force to confront the growing emergence of jihadist fighters. This process entails a stronger level of hierarchical organization and the institutionalization of daily practices, at both the military and civilian levels, that will probably never be needed in a less chaotic context. These soldiers were working to both manage and defend Kobane and its surroundings, sometimes with similar tasks or overlapping duties with the security and political apparatus. Commander Diane confirmed this when he defined his unit as *"soldiers for the government of Kobane and the protection of the border."*[3]

This determined the need for a very structured division of duties and a continuous mobilization of the local Popular Committees that are still working simultaneously as service providers and self-defense groups. For instance, we met groups of armed and unarmed men and women at the crossroads of small alleys, who were helping ordinary citizens to reach their homes during the night due to the almost complete lack of electricity. These are the *heremi*: non-professional volunteers in plain clothes, involved in Popular Committees and protecting their villages, towns, or neighborhoods. Meanwhile, the YPG/YPJ are male and female professional fighters in military uniforms engaged in armed attacks to protect their borders. *Asayş* is the internal security or police, comprised of both men and women, in military uniforms, separate from the YPG/YPJ:

> First of all, there are the local self-defense units (heremi), *then professional fighters and finally the resistance units. Often, men leave the self-defense units to join the YPG/YPJ.*[4]

As with many female fighters, Commander Rangin was not involved in the Popular Committees or within the internal police units (*Asayş*) before joining the YPJ:

> I joined the YPJ initially at a professional level. Many others began as heremi.[5] "I did not work within the Popular Committees before joining the YPG/YPJ".[6] I was not a heremi before joining the YPJ.[7]

The actual experiences of the other fighters within this group were different, as confirmed by our interviewees. Some participants in this focus group never reached professional status, and they still formed part of this unit as local self-defense volunteers,[8] while others were *heremi* before joining the YPG/YPJ[9]:

> *Not all of us will be YPJ for all their lives. I began as a person of my neighborhood defending and controlling our area* [heremi].[10]

On the other hand, Commander Diane was involved in the Popular Committees before joining the YPG: "*I have been working for Rojava (South Kurdistan) before joining the YPG one year ago. I was a* heremi."[11] Diane defined himself as a revolutionary: "*I am ready to fight everywhere. I am a man of the revolution.*"[12] Given his long training with the PKK's commanders in Lebanon, he defended the KPP and its principles: "*The PKK fought for land and freedom. They fight for the concept of freedom and the human being.*"[13]

As for the composition and age of this YPG/YPJ unit's members, it is mixed in terms of ethnic backgrounds,[14] although Kurdish fighters have often been accused of excluding other ethnic groups (Graeber 2017). These soldiers are very young, both men and women, and unmarried. After joining the YPG/YPJ, they cannot marry each other. This is not the case at the *heremi* level:

> *There are both Kurds and Arabs. We fight together [men and women]. Everything happens in friendship, but the soldiers cannot marry*"[15] "*As for women fighters, to be part of the YPJ is more important than a marriage. However, the local components [heremi], as non-professional self-defense groups, are often married people.*[16]

Especially within the female units, the evolution of these self-defense groups has not necessarily led to the formation of a conventional army:

> *We are like every other army; we depend on the ideology of Abdullah Öcalan. But we are not only an army. We are a defensive army.*[17]

Women as fighters

The female fighters (YPJ) appeared to be very well trained by their supervisors, as confirmed by Commander Rangin:

> *The more educated women often enter directly as professional combatants. In meetings, we spend time discussing and self-criticizing. To fight, women must know why and for what to fight. For this reason, we begin with ideological and academic preparation, because every YPJ fighter must know her own self.*[18]

However, according to YPJ's unit commander, there are more men than women working as local volunteers (*heremi*): "*There are more men than women in the first level of engagement. Women usually join at a professional level.*"[19] Thus, there is a very noticeable women's participation, compared to Popular Committees in other MENA countries, within both self-defense groups and resistance units. They are all imbued by feminism and are highly politicized (Knapp et al. 2016). As Commander Rangin explains:

> *We stand for feminism. We depend on ourselves and benefit from the experiences of everyone. Women at home protect the essentialness of women. Our fight is as women (no matter if Kurdish, Syrian, or European) and for a nationality that identifies with democratic autonomy and is opposed to the concept of state. During the fighting for Shingal, for example, women went to save other women. At Til Temir, YPJ fighters went to save Arab women. We went to save dozens of women captives in villages occupied by ISIS.*[20]

Equality between men and women fighters is an essential part of the political formation of these fighters as much as their sexual awareness:

> *Love is essential; it is part of everyone's instinct. The philosophy of death is a way of living. In past times, everyone knew death could come quickly; now it is different, and this disconnects us from nature and does not allow us to accept the idea of death. Religion exploits death: if you are a martyr, you go to heaven. For us, love and death are in contradiction.*[21]

This discourse is closely related to a military and communitarian lifestyle: "*When we discuss this, it is to search for a new military, communitarian, and quotidian life. Women are not made to have only children. We want to reform and renew the community. Thus, we also talk a lot about sexuality*" (Acconcia 2015). According to Rangin, this approach does not result in conflicts with the hierarchy or with male commanders:

> *Some men do not accept that their commanders may be women. If, in this context, the women are soldiers, it is not in vain. We have to fight against the concept that many male comrades have of women. When we talked about this with a YPG soldier, it often happened that he changed his mind and understood that the men's units exist because the YPJ exists and not vice versa.*[22]

As confirmed by our interviewees, the main difference between men and women fighters, as YPG/YPJ soldiers, is that the former appeared to be more educated, with some of them having completed secondary school or wanting to study at university after leaving the YPJ.[23] The YPJ's headquarters in Kobane were very well organized and clean compared to the male units. Some of these women fighters were previously married and later decided to join the YPJ:

Sometimes we are forced to refuse the request to join the female fighters' unit because some women wanted to leave their children alone to be part of our group"[24] Within the local Women's Houses [Mala Jin], there are crèches and other facilities for the children of the fighters[25] Many couples, both men and women, are fighting in their respective units while there are families in which men are ordinary workers and women are fighters.[26]

Moreover, women wearing the *hijab* are also accepted: "*If she [a potential YPJ fighter] is convinced to wear it, we do not complain. Many YPJ fighters wear the* hijab. *In the Commanders' Council, there are women wearing the* hijab."[27]

Thus, women are key to understanding the peculiar system of both political management and security defense in Kobane. As Commander Rangin highlights:

> *The YPJ is not a military auxiliary. Many of our female fighters have been blown skywards by mines; they are commanders (the majority of them) of male units. There is plenty of autonomy regarding this. We have mixed battalions; in almost all battalions, there are co-commanders. In every army, men attack without considering the values of this community, with women as fighters giving up doing so. For instance, if Kurdish fighters do not commit ethnic cleansing after the capture of a city, this is mainly because our influence stops errors from being committed.[28]*

From Popular Committees to fighting units

In this section, the developments within the organization of the initial mobilized self-defense groups in Northern Syria between 2011 and 2016 will be described. In other words, I will further disentangle the process of mobilization of the local Popular Committees' participants, who often begin as volunteer self-defense civilians, later transforming themselves into YPG/YPJ professional fighters in Kobane, which forms part of the discussion in the interviews in our focus group.

Kobane lived in a context of the constantly high mobilization of ordinary people between 2011 and 2016. However, a standardized system of recruitment and career paths, within both Popular Committees and YPG/YPJ units, has not been established. A standard period of compulsory training is usually required:

> *The system is not stabilized yet. There are many exceptions, especially in periods of general mobilization. However, when local men or women become* heremi, *they usually need at least three months to decide if they want to continue both as self-defense volunteers and professional fighters"[29] "We do not sign any contract to be part of the YPG/YPJ.[30]*

However, it is still problematic to state exactly at what stage of the Syrian War these groups changed their nature and transformed themselves from nonviolent defenders into armed defensive units:

> *At the beginning of the uprisings, we did not carry weapons. We depended only on the support of ordinary people, not on weapons.*[31]

This often happened when the Syrian regime withdrew from the northern regions between 2012 and 2013. During this period, YPG/YPJ commanders officially announced their armed struggle. As Commander Rangin recalls:

> *After the liberation of northern Syria by the Syrian regime, we took many weapons. Everything changed. At the end of 2012, we announced the beginning of our armed struggle. Later, in the Jazira and Kobane provinces, we seized weapons. Only recently [2014], we received a number of Kalashnikovs from abroad, while some foreigners joined our units.*[32]

In the initial engagement of local volunteers in self-defense groups, the concept of legitimate defense was vital:

> *If someone attacks you, you have to attack him or her for legitimate defense, positive or active defense. If the attacks intensify, we intensify the attacks, too. The target is to protect ourselves before the enemy attacks us.*[33]

This seems to be not only a military aim, but also a political one for these YPG/YPJ members. For example, as I witnessed, when those soldiers liberated Tall Abyad from ISIS (2015), they lifted the flag of their army as a symbol of both political and military power in Kobane Canton:

> *Sometimes we attack militarily, but we resist politically. The political and military defenses are overlapping.*[34]

Mobilization and strategy

The majority of the interviewees were politically or emotionally involved in the Syrian civil war before joining the YPG/YPJ. It seems that if, at an initial stage, they joined local committees only with the aim of protecting their homes from a lack of security, following the emergence of ISIS and its permanent occupation of Kobane, the same individuals became highly motivated to be part of the armed struggle. Some were motivated by the killing of a relative by ISIS[35]:

> *I was a member of the democratic youth of PYD before joining YPG. My mother is working for a local institute supporting martyrs' families"*[36]

*One of my brothers is a martyr"[37] My cousin was killed by ISIS"[38]
I came from Lebanon to Syria when I knew about the ISIS attacks"[39]
I was working for the PYD before joining the YPG.[40]*

The main reason to be part of the fight against the jihadists appeared to be
to protect their land from ISIS:

*ISIS is our biggest problem"[41] We fight to protect our land[42] We fight
against ISIS, to free our land and for our rights[43] This does not mean
to fight only for the Kurds but for the freedom of everybody[44] I fight
because I want my family back. They escaped to Turkey after the ISIS
attacks.[45]*

However, this has nothing to do with religious motivation:

*They [ISIS] do things in the name of Islam, but this is not right. No
religion says to shoot another man[46] They are not humans; they are
monsters. They do not represent Islam.[47]*

Nor did the Assad regime have a better reputation among these YPG/YPJ
soldiers, although they were more careful in their judgments: "*He [Assad]
was controlling us.*"[48]

As for the female fighters, they seemed to have as their first objective the
defense of their own people and of women in more general terms:

*We fight to protect our people"[49] I fight for the Kurdish people and the
martyrs.[50]*

Many of those YPJ fighters considered ISIS and Turkey to be allies:

*The jihadists are monsters and Turkey had much responsibility for help-
ing them[51] ISIS is the first enemy of women[52] In Shingal, the jihadists
were so aggressive against women; it is our duty to fight them.[53]*

As Polda, a YPJ soldier, added:

*After three months of training, I can fight everywhere in the region
(Rojava). I am fighting for the freedom of my land.[54]*

Many of these fighters felt abandoned by their so-called foreign allies. For
instance, the commander of this YPG's unit did not consider the support
given at that stage by the airstrikes of the US-led coalition as effective. As
Commander Diane stated,

*If the coalition would like to bomb a cigarette, they do. But if we ask to
bomb an ISIS target, often they do not listen to our requests.[55]*

However, at the lower level, the interviewees considered the US strikes helpful while highlighting that the real enemies were the Turks:

> *They [the US-led coalition] are helping us*[56] *Our [Kurds'] biggest enemies are the Turks*[57] *Turkey supports ISIS. They opened the borders to let ISIS fighters escape when we attacked them*[58] *The US coalition is helping, but they could do much more.*[59]

As for the use of weapons, according to Commander Diane at the time of the focus group, there was a daily need to use weapons against ISIS fighters by his unit's components:

> *They [ISIS] attack every day even if they do not have great experience and are gradually losing their military capacity.*[60]

However, the YPG soldiers' engagement did not end with the liberation of Kobane from ISIS in 2015:

> *During and after the liberation of Kobane, we had to free and control the town house by house, road by road. The jihadists were always 10 miles from us. Now, we are involved in the liberation of other parts of the canton in order to connect it with Jazira Province.*[61]

Finally, some of those fighters appeared to be willing to continue as civilians, and others as combatants once the situation had been stabilized:

> *I will be a combatant after the war, too. I will work for my town and my country*[62] *I joined YPJ, but this will not continue until the end of my life, as I am not a professional soldier.*[63]

Nonetheless, all of them will pursue the defense of their political behavior as imbued by Öcalan's books and experiences:

> *Only Öcalan's philosophy brought us to this level. He gave us the opportunity to know who we were. He made us understand our culture*[64] *Öcalan's theory is for all nationalities.*[65]

Thus, it remains unclear whether, in a more stabilized environment, these fighters will choose to return to their civilian activities, to contribute to institutionalizing a permanent armed wing of the PYD, or to be integrated into the army and security personnel of a potentially autonomous state.

Conclusion

Left-wing activism was boosted by the 2011 uprisings in the MENA region. If we look from a bottom-up perspective at the uprisings, although the ongoing social movements had no strong international connections both within the region and worldwide, there are many commonalities between the patterns of mobilization and demobilization within mobilized groups between these countries.

This chapter has focused on the Syrian Popular Committees because, in both Egypt and Northern Syria, the initial grassroots mobilization challenged the traditional patterns of state control over civil society. Later, in the context of the war in Northern Syria, the participants within those Popular Committees were found to feel the compulsion to engage in an armed struggle in order to protect their neighborhoods and fight against the growing emergence of jihadist fighters.

Thus, in Syria, these social movements evolved into paramilitary organizations; in other words, they were paramount in forming armed entities like the YPG-YPJ. This clearly surfaced when the Syrian regime withdrew from the northern regions of the country in 2012, and those units' commanders officially announced their armed struggle.

This process determined a stronger level of hierarchical organization and the ongoing institutionalization of daily practices, at both the military and civilian levels, with very noticeable women's participation, within both self-defense groups and the resistance units.

If, at a preliminary stage, the interviewees joined local committees with the primary aim of protecting their homes due to a lack of security, they later became highly motivated to be part of the armed struggle, especially with the emergence of ISIS fighters and their permanent occupation of Kobane.

Examining the Popular Committees as a long-term and politicized phenomenon at the urban level in Northern Syria has allowed a better understanding of the process of control over the mass uprisings sparked in response to state repression of peaceful protest by these emergent grassroots organizations, as well as the second phase of growth of hierarchical structures in response to the outbreak of full-scale war in Syria.

Notes

1 The PKK is included in lists of designated terrorist groups by the US and the UAE.
2 See also https://bianet.org/english/freedom-of-expression/165514-three-italian-and-a-french-journalist-deported [Last accessed March 1, 2017].
3 Interviewee 1.
4 Interviewee 2.
5 Interviewee 2.
6 Interviewee 11.
7 Interviewee 12.
8 Interviewee 7.
9 Interviewees 3, 4, 6, and 8.
10 Interviewees 13 and 14.
11 Interviewee 1.
12 Interviewee 1.
13 Interviewee 1.

14 In October 2015, the Syrian Democratic Forces (SDF) were founded. They represent a multi-ethnic and multi-religious alliance that includes Kurds, Arabs, Assyrian, Armenian, Turkmen, and Circassian fighters.
15 Interviewee 1.
16 Interviewee 2.
17 Interviewee 2.
18 Interviewee 2.
19 Interviewee 2.
20 Interviewee 2.
21 Interviewee 2.
22 Interviewee 2.
23 Interviewee 2.
24 Interviewee 2.
25 Interviewee 2.
26 Interviewee 2.
27 Interviewee 2.
28 Interviewee 2.
29 Interviewee 2.
30 Interviewee 3.
31 Interviewee 2.
32 Interviewee 2.
33 Interviewee 2.
34 Interviewee 2.
35 Interviewees 5, 9, and 14.
36 Interviewee 7.
37 Interviewee 4.
38 Interviewee 5.
39 Interviewee 3.
40 Interviewee 11.
41 Interviewees 3, 4, and 5.
42 Interviewees 5, 6, and 7.
43 All interviewees.
44 Interviewee 3.
45 Interviewee 3.
46 Interviewees 4 and 5.
47 Interviewees 6 and 7.
48 Interviewees 8 and 9.
49 Interviewees 11 and 12.
50 Interviewees 13 and 14.
51 Interviewees 12 and 13.
52 Interviewee 11.
53 Interviewee 12.
54 Interviewee 14.
55 Interviewee 1.
56 Interviewee 8.
57 Interviewee 9.
58 Interviewee 9.
59 Interviewee 12.
60 Interviewee 1.
61 Interviewee 1.
62 Interviewees 13 and 14.
63 Interviewees 12 and 13.
64 Interviewees 4, 5, 6, 11, and 15.
65 Interviewees 11 and 14.

References

Acconcia, G. (2015). Il Kurdistan non è vicino. Le guerre islamiche. *Limes, 9*, 181–188.

Allsopp, H. (2015). *The Kurds of Syria: Political parties and identity in the Middle East*. London: I. B. Tauris.

Allsopp, H., & Acconcia, G. (2013). I kurdi siriani, né con al-Assad né con le opposizioni. *Il Manifesto*, October 1.

Bookchin, M. (2015). *The next revolution: Popular assemblies and the promise of direct democracy*. London: Verso.

Boothroyd, M. (2016). Self organisation in the Syrian revolution. *Socialist Project*, October 10.

El-Meehy, A. (2017). Governance from below. Comparing local experiments in Egypt and Syria after the uprisings, report Arab politics beyond the uprisings. *The Century Foundation*.

Gelvin, J. (1994). The social origins of popular nationalism in Syria: Evidence of a new framework. *International Journal of Middle East Studies, 26*(4), 645–661.

Gelvin, J. (1998). *Divided loyalties. Nationalism and mass politics in Syria at the close of Empire*. Berkeley, CA: University of California Press.

Graeber, D. (2017). Why is the world ignoring the revolutionary Kurds in Syria? *The Guardian*, March 1.

Hassan, H. (2015). Extraordinary politics of ordinary people: Explaining the micro dynamics of popular committees in revolutionary Cairo. *International Sociology, 30*(4), 383–400.

Knapp, M., Flach, A., & Ayboga, E. (2016). *Revolution in Rojava. Democratic autonomy and women's liberation in Syrian Kurdistan*. London: Pluto Press.

Conclusion

In this book, we applied social movement theories to the MENA region in order to study the so-called Arab Spring. We focused mainly on the Egyptian, Tunisian, and Syrian case studies to disentangle the political and organizational micro-dynamics of the upheaval, in both urban and peripheral areas, to acknowledge the reasons that determined the impossibility to forge cross-ideological coalitions between socialists and Islamists in the aftermath of the upheaval, to compare Egypt and Syria in terms of grassroots mobilizations, and to analyze the 2011 uprisings as a "leftwing awakening" and a fight for women's rights, rather than as an "Islamic awakening," as they have been often portrayed.

The protests that took place in the public space in these countries were part of a longstanding and already normalized struggle against state repression perpetrated for decades in these authoritarian regimes. In particular, alternative forms of contentious actions emerged as a number of fragile and diverse social movements were able to mobilize a growing number of activists and ordinary citizens.

Leftist activists in the MENA region forged or reinvigorated unprecedented movements, established trade unions, or political parties as an effect of the 2011 uprisings: the EFITU in Egypt; UGTT in Tunisia; the People's Democratic Party (HDP) in Turkey; and the Democratic Union Party (PYD) in Syria. Thus, in our case studies, we argued that these have been the most relevant outcomes of the recent grassroots mobilizations, despite the prominent mainstream relevance given to moderate or radical Islamist groups, initial movements, or political parties.

This happened because the 2011 uprisings took place in a context in which the disenfranchised and the working classes were especially significant and had long been neglected by formal politics. In other words, in 2011, these classes began to be more structured and organized in Egypt, Tunisia, Syria, and Turkey. However, despite their popularity, they have often not been well integrated in urban proto-movements (Egypt and Syria) or in traditional parliamentary politics (Turkey and Tunisia).

For the Egyptian workers' movements, we argued that they appeared to be too fragmented, leaderless, and lacking in a coherent ideology. However,

DOI: 10.4324/9781003293354-8

in comparison with the period before the uprisings, the EFITU seemed to tackle the demands for workers' rights with a more independent approach from the government's economic policies. However, the trade unions within the EFITU, as well as other independent trade union umbrellas, have never been clearly legalized by the Egyptian authorities, and they have often been used by state security agencies to control mobilized workers.

In Tunisia, the workers' movements succeeded. This was especially due to a more compliant political attitude, as shown by the UGTT, whose role may have been one of the most important factors that enabled the emergence of a more inclusive coalition compared to other countries. Thus, the UGTT represented the "heart of the coordination between the diverse actors of the protests." Compared to the governmental trade unions in Egypt, the UGTT in Tunisia gained autonomy from the state. This was vital in mobilizing the proto-movements, especially in peripheral areas, and in the formation of a comprehensive coalition that had as its first objective a comprehensive democratic transition. However, even though the UGTT has been more effectively involved in the Tunisian transition toward a more democratic political system, this process has still failed to advance an effective representation of labor rights.

The political involvement in Turkish party politics and parliament of the HDP is another good example of a recent left-wing social movement in the region that grew in the aftermath of popular uprisings (in this case, the 2013 Gezi Park movement). Thus, the grassroots mobilizations in Turkey had as one of their most significant outcomes the strengthening of the pro-Kurdish Left-communalist party (the HDP). In other words, despite state repression, the grassroots mobilizations had enough time to organize and structure their opposition to the traditional party political system and channel their grievances into an effective political party with a notable constituency, albeit with only limited political impact.

The same can be said of Northern Syria, where the left-wing Democratic Union Party (PYD) emerged as the most important political group to lead the transition process in this region, largely thanks to the support given by the YPG/YPJ militias. These fighting units are evolving into more comprehensive forms of paramilitary groups, respectful of gender and ethnic minorities (SDF, the Syrian Democratic Forces): "*We are fighting, and we are dying for the freedom of this land. We will continue to resist.*" These were the last words of Viyan Qamishlo, a young YPJ fighter killed in Manbji in September 2016 during clashes between the Syrian Kurds and ISIS. However, at the time of writing, the revolutionary dream of an autonomous Rojava in Northern Syria has again been significantly challenged by the advance of the Turkish army, which has militarily occupied Afrin, undermining the territorial continuity of Rojava in the context of the announced withdrawal of US soldiers based in Northern Syria.

Particularly with regard to the Rojava case, when we use a gendered and intersectional perspective to look at the effects that the war has on the lives of individuals, the issue becomes much more complicated, and the micro-level of analysis that we have chosen through the use of in-depth interviews sheds new light not only on the fact that, as a side effect, we have the cancellation of the basic human rights for all, but also on a misleading construction of subjectivities, so that women in the aid system are only seen as victims, women Kurdish fighters are only heroines sacrificed to the battle, and LGBT people are another vulnerable target. We continue not to give agency to these people, to speak for them, to decide what their true problems are, and what they need to resolve them.

In more general terms, in this book, we argue that the 2011 uprisings in the MENA region showed a certain incompatibility between Political Islam and the Left. This was apparent in at least three of our case studies: in Egypt between the Muslim Brotherhood and the Left, and in Turkey and Northern Syria with the repression of the HDP, the PKK, and the PYD carried out by the Turkish authorities, especially after the 2016 failed military coup. Thus, we concluded that in postrevolutionary settings, counterrevolution may take place when revolutionary coalitions are not able to build long-lasting alliances that cut across the specific interests and peculiarities of the single groups. In Egypt, old regime actors capitalized on a divided political field to undermine alliances between secularists and Islamists, ultimately recapturing state institutions. In Tunisia, however, the revolutionary coalition, rooted in prerevolutionary settings, lasted longer, and included both secularist and Islamist actors. It was then able to promote conciliation, leading to the new 2014 Tunisian Constitution. Democratic transition, in this case, also occurred thanks to the capacity of civil society organizations to engage with a variety of actors, including political parties. This highlights the positive role of popular uprisings in supporting processes of democratization.

Ten years after the 2011 uprisings in the Middle East, the space for social and political change has narrowed due to strengthened military regimes and ongoing civil wars. Authoritarian regimes have reinvigorated a populist and xenophobic discourse, reproducing the same kind of fear and mistrust justified by anti-terrorism rhetorical policies in other MENA countries. In this context, in Egypt and Syria, Abdel Fattah al-Sisi and Bashar al-Assad have consolidated their power, while in Turkey, Recep Tayyip Erdogan has reduced the space for political dissent.

In both Tunisia and Egypt, subalterns were paramount in the formation of popular mobilizations calling for human and social rights, minimum wages, and social justice. Key actors in the 2010–2011 uprisings in Tunisia have been, on the one hand, workers engaged in major trade unions like the UGTT, which have been central in the local protests that occurred long before 2011 (Beinin & Vairel 2011) and, on the other hand, the ATFD. Not only has the ATFD played an important role in the opposition to the regime throughout the 1990s and the new millennium, but it has also focused on gender rights against Ben Ali's state feminism, Islamism, and rising conservatism (Debuysere 2018). Similarly, protests erupted in Egypt in January 2011 as the product of large cross-class networks in which young people and students joined middle-class professionals, government employees, workers, housewives, and the unemployed. These mass riots were paramount in the formation of new means of popular mobilization aimed at enhancing a diverse range of unmet needs and motivating ordinary citizens to participate in socio-political activities, such as advocating for human and social rights, calling for minimum wages and workers' rights, security, and self-defense, and participating in the political arena. In the aftermath of the uprisings and until early 2013, free elections were held in both Egypt and Tunisia, assemblies were charged with drafting a new constitution, and Islamist parties won elections and assumed office (Hassan et al. 2020). After 2013, Egypt and Tunisia show divergent profiles. In Tunisia, a democratic transition culminated in the adoption of a new Constitution on January 14, 2014, with pertinent achievements for Tunisian women, such as the approval of several laws increasing women's political and legal rights, including Law 58, which has criminalized violence against women since 2017.

Throughout this book, we have tried to show how difficult it is to inscribe women's participation in the construction of mobilization scenarios. Not all of them are easy to describe, and not all of them fall within the usual canons of participation. The instruments we have chosen – interviews and focus groups – have the power to give a voice back to people, to support and enhance their demands, and to support their mobilization without necessarily framing it with our usual white and western eyes. Mobilization can take the shape of a resistance, a hiding at the margins, or an open participation in the battle. Whichever the case, we are not allowed to translate the message but just to make it heard.

References

Beinin, J., & Vairel, F. (Eds.). (2011). *Social movements, mobilization, and contestation in the Middle East and North Africa*. Stanford, CA: Stanford University Press.

Debuysere, L. (2018). Between feminism and unionism: The struggle for socio-economic dignity of working-class women in pre- and post-uprising Tunisia. *Review of African Political Economy*, 45(155), 25–43.

Hassan, M., Lorch, J., & Ranko, A. (2020). Explaining divergent transformation paths in Tunisia and Egypt: The role of inter-elite trust. *Mediterranean Politics*, 25(5), 553–578. doi: 10.1080/13629395.2019.1614819

Index